The Mechanics
of Inheritance

PRENTICE-HALL FOUNDATIONS OF MODERN *Genetics* SERIES

Sigmund R. Suskind and Philip E. Hartman, Editors

AGRICULTURAL GENETICS
    James L. Brewbaker

GENE ACTION, Second Edition
    Philip E. Hartman and Sigmund R. Suskind

EXTRACHROMOSOMAL INHERITANCE
    John L. Jinks

DEVELOPMENTAL GENETICS*
    Clement Markert and Heinrich Ursprung

HUMAN GENETICS, Second Edition
    Victor A. McKusick

POPULATION GENETICS AND EVOLUTION
    Lawrence E. Mettler and Thomas G. Gregg

THE MECHANICS OF INHERITANCE, Second Edition
    Franklin W. Stahl

CYTOGENETICS
    Carl P. Swanson, Timothy Merz, and William J. Young

*Published jointly in Prentice-Hall's *Foundations of Developmental Biology Series.*

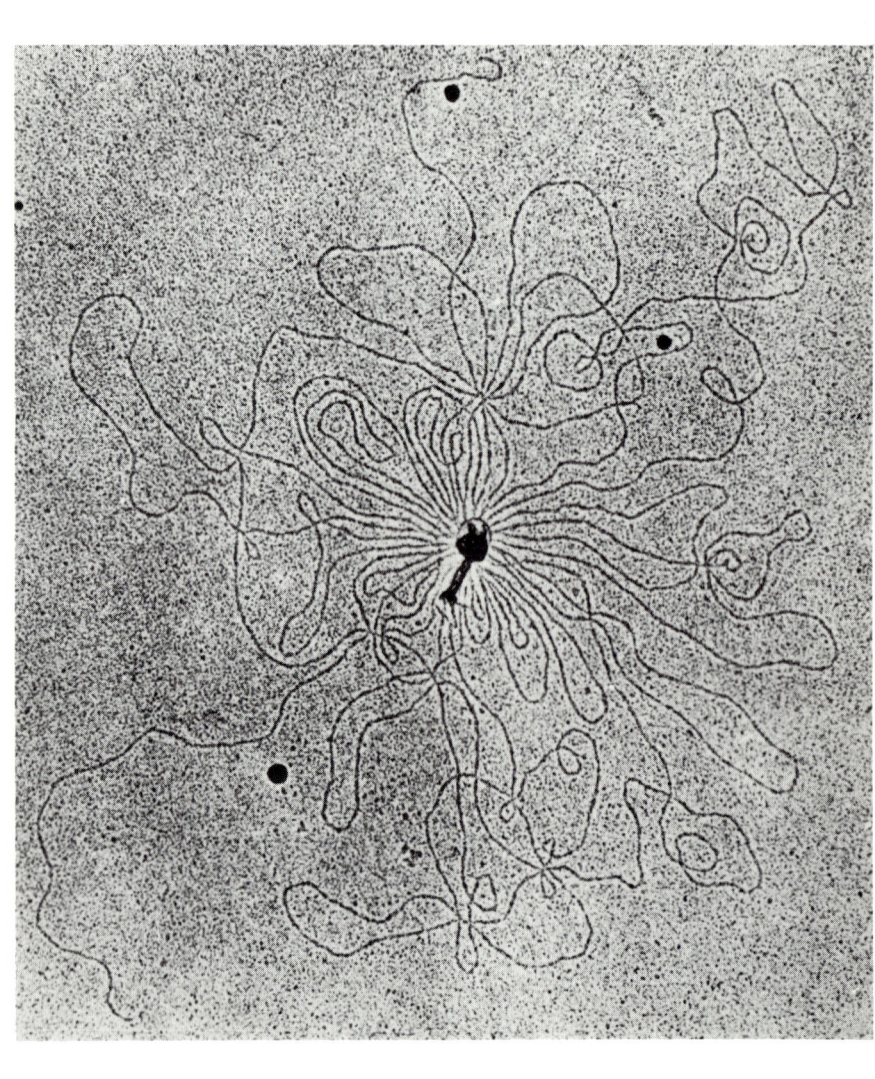

THE MECHANICS OF INHERITANCE

Second Edition

Franklin W. Stahl
University of Oregon

PRENTICE-HALL, INC. *Englewood Cliffs, New Jersey*

FOUNDATIONS OF MODERN GENETICS SERIES
Sigmund R. Suskind and Philip E. Hartman, Editors

© *Copyright 1964, 1969*
by PRENTICE-HALL, INC.
Englewood Cliffs, New Jersey
All rights reserved. No part of this book
may be reproduced in any form,
or by any means without
permission in writing from the publisher.

Printed in the United States of America

C-13-571059-6
P-13-571042-1
Library of Congress Catalog Card Number:
69-19870

Current printing (last digit):
10　9　8　7　5　4　3　2　1

FRONTISPIECE. *The protein coat of a bacterial virus lying on the DNA experimentally released from it. The photograph, kindly supplied by D. Lang, appeared in* Biochim. Biophys. Acta, 61 *(1962), 857.*

PRENTICE-HALL INTERNATIONAL, INC., *London*
PRENTICE-HALL OF AUSTRALIA, PTY. LTD., *Sydney*
PRENTICE-HALL OF CANADA, LTD., *Toronto*
PRENTICE-HALL OF INDIA PRIVATE LTD., *New Delhi*
PRENTICE-HALL OF JAPAN, INC., *Tokyo*

575.1
St14m
6035

For Mary, Emily, Josh, and Andy

Foundations of Modern *Genetics*

The books in this series are intended to lead the alert reader directly into the exciting research literature of modern genetics. The forefront of genetic research draws heavily on concepts and tools of chemistry, physics, and mathematics. Because of this, the principles of genetics are presented here together with discussions of other relevant scientific areas. We hope this approach will encourage a fuller comprehension of the principles of genetics and, equally important, of the types of experiments that led to their formulation. The experimental method compels the questions: What is the *evidence* for this concept? What are its *limitations*? What are its *applications*?

Genetics today is penetrating increasingly into new areas of biology. Its rapidly expanding methodology is enabling research workers to find answers to questions that it was futile to ask only a short while ago. Even more provocative studies now underway are raising new, heretofore unimagined questions.

The design of the individual short volumes of the Prentice-Hall Foundations of Modern Genetics Series permits stimulating, selective, and detailed treatments of each of the various aspects of the broad field of genetics. This facilitates more authoritative presentations of the material and simplifies the revisions necessary to keep abreast of a rapidly moving field. Each volume has its own individual focus and personality while at the same time overlapping with other volumes in the Series sufficiently to allow ready transition. Collectively, the Series, now complete, covers the main areas of contemporary genetic thought, serving as a thorough, up-to-date textbook of genetics—and, we hope, pointing the reader toward experiments even more penetrating than those described.

SIGMUND R. SUSKIND
PHILIP E. HARTMAN

Preface to the Second Edition

Only five years have elapsed since the appearance of the first edition of *The Mechanics of Inheritance*. However, when the Editors suggested that it was due for a thorough revision, I agreed enthusiastically. Recent developments in genetics were of such magnitude as to make the revision necessary. Equally compelling was the awareness of various omissions in the first edition that became apparent to me through use of that book in several courses at the University of Oregon.

This new edition employs the strategy of the old one; were it to do otherwise, it would need a new author and a new title. However, within that framework, substantial changes and additions have been effected. I hope these modifications have rendered the facts of inheritance and current ideas about them both more accessible and more interesting to the student at all levels.

R. W. Siegel, Philip E. Hartman, and the students of Biology 104, 301, and 320 have contributed comments especially valuable in the preparation of this revision.

F.W.S.

Contents

One   Heredity   1

MICROORGANISMS IN GENETIC RESEARCH   2
THE GROWTH OF BACTERIAL CULTURES   3
ENZYMES   7
MUTATION IN BACTERIAL CULTURES   8
MEASUREMENT OF MUTATION RATE   11
MUTATIONS AND ENZYMES   14
SUMMARY   16
REFERENCES   16
PROBLEMS   16

Two   The Genic Material   19

BACTERIAL TRANSFORMATION   19
IDENTIFICATION OF THE TRANSFORMING AGENT   23
THE LIFE CYCLE OF BACTERIOPHAGE T2   23
RNA VIRUSES   27
THE STRUCTURE OF DNA   27
THE RARE DEOXYRIBONUCLEOTIDES   37
THE STRUCTURE OF VIRAL RNA   38
SUMMARY   40
REFERENCES   40
PROBLEMS   41

Three   Duplication of Nucleic Acid   42

"SEMICONSERVATIVE TRANSMISSION" OF ISOTOPIC LABEL   43
IN VITRO SYNTHESIS OF GENIC NUCLEIC ACID   49
SUMMARY   51
REFERENCES   51
PROBLEMS   51

xiv   THE MECHANICS OF INHERITANCE

Four   Mutation of DNA   54

TAUTOMERISM OF BASES   *54*
MUTATION BY BASE-PAIR TRANSITIONS   *59*
BASE-PAIR TRANSITIONS INDUCED BY 5-BROMOURACIL AND
    NITROUS ACID   *62*
MUTATIONAL HETERODUPLEXES   *65*
SEGREGATION FROM HETERODUPLEXES   *66*
OTHER MUTATIONS IN DNA   *67*
MUTAGENIC MUTATIONS   *68*
SUMMARY   *69*
REFERENCES   *69*
PROBLEMS   *70*

Five   Organization of Genic Material   72

PACKING OF DNA   *74*
VIRUS CHROMOSOMES   *76*
THE BACTERIAL CHROMOSOME   *78*
CHROMOSOMES OF HIGHER ORGANISMS   *81*
ORGANIZATION OF DNA WITH RESPECT TO DUPLICATION   *88*
SUMMARY   *91*
REFERENCES   *91*
PROBLEMS   *91*

Six   Recombination in Higher Organisms: I   94

MEIOSIS   *95*
SEGREGATION   *99*
RECOMBINATION   *100*
CROSSING OVER   *101*
THE MAPPING FUNCTION   *102*
DEPARTURE FROM THE IDEAL MAPPING FUNCTION   *105*
THREE-FACTOR CROSSES   *106*
THREE-FACTOR CROSSES   *106*
SUMMARY   *107*
REFERENCES   *108*
PROBLEMS   *108*

Supplement   Meiosis in Salamander Spermatocytes   112
by James Kezer

SUMMARY   *130*

Seven   Recombination in Viruses   131

A PHAGE CROSS   *131*
COMPARISON OF A PHAGE MAP WITH ITS CHROMOSOME   *133*
THE "MECHANISM" OF RECOMBINATION   *139*
FINE-STRUCTURE ANALYSIS   *143*
ORGANIZATION OF IMMATURE ("VEGETATIVE") VIRAL DNA   *146*
SUMMARY   *151*
REFERENCES   *151*
PROBLEMS   *152*

Eight   Recombination in Bacteria   154

PHAGES AND THE BACTERIAL CHROMOSOME   *154*
TRANSDUCTION   *155*
EPISOMES IN BACTERIAL FERTILITY   *161*
THE MECHANISM OF RECOMBINATION   *163*
RECOMBINATION-DEFICIENT MUTANTS   *163*
SUMMARY   *164*
REFERENCES   *164*
PROBLEMS   *164*

Nine   Recombination in Higher Organisms: II   167

ASCOSPORE FORMATION IN NEUROSPORA   *167*
HIGH NEGATIVE INTERFERENCE   *168*
ABERRANT SEGREGATION   *169*
ASYMMETRIC RECOMBINATION   *171*
THE MODEL   *173*
SUMMARY   *179*
REFERENCES   *179*
PROBLEMS   *180*

Ten   The Code   181

THE MECHANISM OF PEPTIDE SYNTHESIS—AN OUTLINE   *182*
FRAME-SHIFT MUTATIONS AND THE GENERAL NATURE OF THE
    GENETIC CODE   *186*
CRACKING THE CODE   *188*
REFERENCES   *191*
PROBLEMS   *191*

Eleven The Genetic Analysis of Diploids 193

GENOTYPE AND PHENOTYPE *193*
PHENOTYPIC LAG AND PHENOTYPIC MIXING *193*
DOMINANCE AND RECESSIVENESS *194*
NATURE AND NURTURE *195*
COMPLEMENTATION *195*
HETEROZYGOSIS IN BACTERIA *195*
DIPLOIDY IN HIGHER ORGANISMS *196*
REFERENCES *197*
PROBLEMS *197*

Appendix The Poisson Distribution 201

REFERENCES *203*
PROBLEMS *203*

Answers to Problems 207

Subject Index 211

Author Index 215

The Mechanics
of Inheritance

Heredity

A common observation is that cats give birth to cats. Cats have been up to this for centuries all the world over. Furthermore, there is no documented instance of a cat giving birth to anything else. This is the primary observation of the science of heredity.

Similarly, Douglas firs produce seed that germinate to give only Douglas firs. The fruit fly, *Drosophila*, has been watched by the world's great geneticists since 1910; it has never produced any other animal than *Drosophila*. Cats and firs are likewise objects of genetic research. They do have the advantage of being prettier than flies, but they share with *Drosophila* two major disadvantages as objects of study. They are big (requiring lots of lab space), and they are slow to reproduce (requiring lots of lab time).

These familiar creatures possess in common another feature—their developmental cycles are sufficiently complicated that an analysis of their heredity requires a simultaneous understanding of *all* the basic principles of genetics. One objective of this series is to provide the basis for such an analysis, and this volume has been written with that goal in mind. The primary aim of this volume, however, is to focus on the physical basis of inheritance, and that goal is better served by emphasizing the revealing experiments on heredity in microorganisms.

1

Microorganisms in Genetic Research

A number of factors have contributed to the success of these experiments. The microorganisms are small (requiring little lab space) and fast (requiring little lab time). In addition, the developmental cycle of each microorganism is sufficiently simple that the basic principles of genetics can conveniently be illustrated one at a time. On the other hand, when considered collectively, microorganisms provide suffi-

FIG. 1.1. *A stereo electron photomicrograph of a pair of bacterial cells engaged in sexual conjugation (see Chapt. 5). In order to distinguish conjugating pairs from pairs of cells that have arisen by division, male and female strains which differ in appearance were permitted to conjugate. One of the strains is piliated (hairy) while the other is not piliated but can adsorb tadpole-shaped virus particles to its surface, and has done so. A three-dimensional view of the whole can be perceived, after a little effort, if you hold the book about two feet from your eyes and stare at the figure for a moment or two. If this doesn't work, try putting one end of a piece of paper between the photomicrographs and the other end against your nose, so that you see only one picture with each eye. This photo was kindly supplied by the photographer, Thomas F. Anderson.*

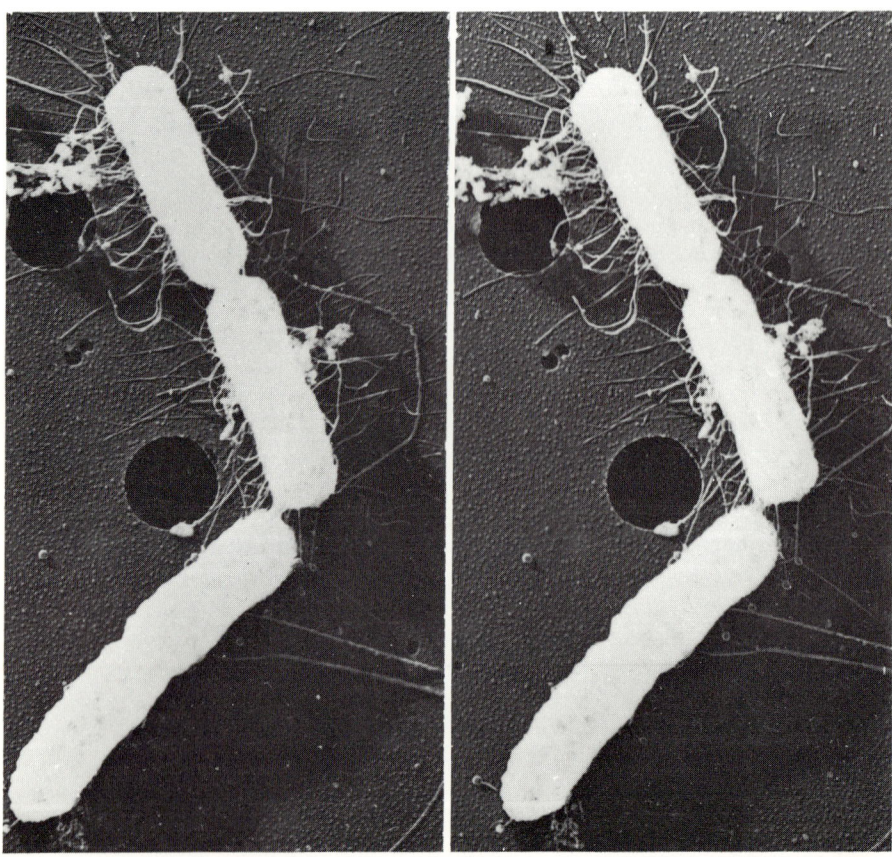

cient diversity to permit step-by-step illustration of the features of more complicated systems.

Let us look at the microorganisms that will carry us through this first chapter (Fig. 1.1).

In 1946, E. L. Tatum wrote (in *Cold Spring Harbor Symposia on Quantitative Biology*): "The bacteria, with their biochemical and physiological versatility, ease of cultivation and study ... may prove excellent material for the study of the fundamental problems of ... genetics. ... The main attribute lacking in bacteria ... is their apparent lack of a sexual phase, the existence of which would permit their examination by classical genetic methods for the segregation of characters as Mendelian units." We shall see in subsequent chapters that bacteria are not only free of this defect, but that, on the contrary, they are superb material for the study of "Mendelian units." At present, we shall entertain them for their earlier recognized virtues.

The Growth of Bacterial Cultures

A bacterium reproduces itself in an apparently simple fashion: it elongates, then splits transversely into two bacteria; after a suitable interval, each of the daughter bacteria reproduces similarly. We can

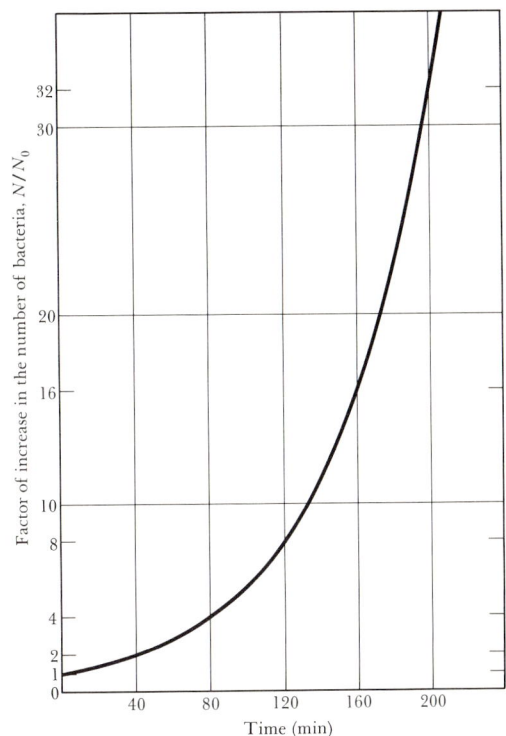

FIG. 1.2. *The increase in the number of bacteria in an actively dividing bacterial culture as a function of time. In an actively dividing culture, the number of bacterial cells increases at a rate proportional to the number of cells present at any moment.*

plot the number of bacteria as a function of time in a culture maintained under constant, favorable conditions (Fig. 1.2). We see that a doubling every 40 min, for instance, leads to a thousandfold increase in about 400 min.

The relationship between the number N of bacteria in a culture and the number of times the population has doubled is written as

$$N = N_0 2^g \qquad \text{(Eq. 1.1)}$$

where N_0 is the number of bacteria present at time zero and g is the number of generations (population doublings) that have occurred. An actively dividing culture maintained under constant conditions multiplies at a constant rate, so that

$$g = kt \qquad \text{(Eq. 1.2)}$$

where t is time and k is a growth-rate constant.

The consequences of this mode and tempo of reproduction are more

FIG. 1.3. *The increase in the number of bacteria in an actively dividing culture plotted logarithmically as a function of time. The three different scales which are commonly employed for such a plot are shown along the ordinate.*

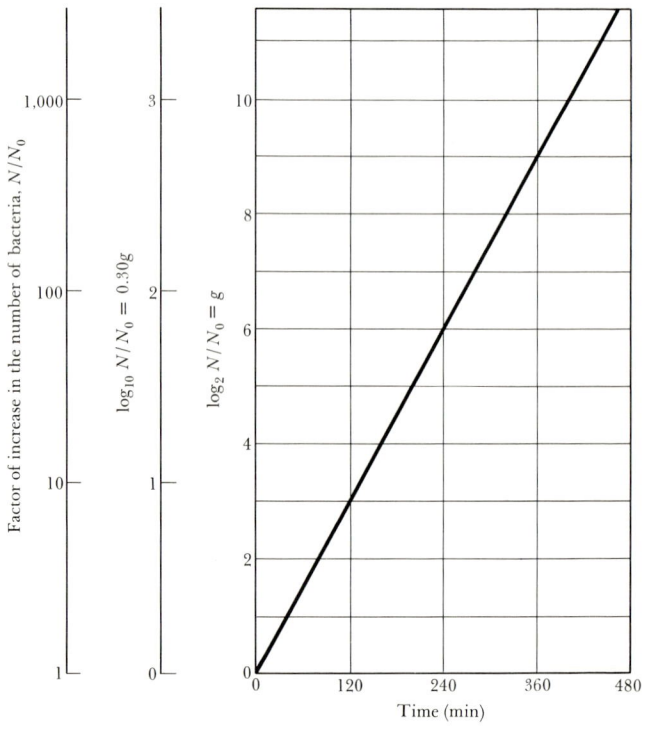

easily seen from a plot of the logarithm of the number of bacteria as a function of time (Fig. 1.3). The linear relationship between time and the log number of bacteria (Fig. 1.3) is easy to extrapolate to times beyond those represented on the graph. When multiplication proceeds at the indicated rate of one doubling each 40 min, after 22 hr the initial bacterium has given rise to a population of greater than ten billion individuals, more than three times the number of the human population of the earth.

The bacterium *Escherichia coli* is the type most exploited in the genetics laboratory. It displays the simple growth habit described above, but when subjected to chemical analysis it proves to be about as complicated as any creature. The principal classes of compounds found in *E. coli* are listed in Fig. 1.4b. This array of chemicals, which characterizes each of the billions of individuals in a culture, is to be contrasted with the handful of simple compounds (Fig. 1.4a) that must be present in the culture medium in order for *E. coli* to duplicate. These few compounds provide sources of each of the chemical elements out of which a bacterium is composed and, in addition, a source of energy to drive synthetic reactions. *E. coli* (like any other organism) can arrange the atoms in the compounds available to it into a biochemical and morphological likeness of itself, and it does so at a rate that permits it to duplicate once every 40 min. The myriad chemical conversions carried out by *E. coli* (or any other creature) can proceed at high speed because they are catalyzed by enzymes.

Enzymes

Escherichia coli contains a couple of thousand (my guess) different kinds of enzyme molecules. Each of these kinds catalyzes a particular step in the degradation of a carbon source (such as glucose) and the utilization of the resulting fragments in the synthesis of the various compounds which make up the cell. The energy released from a number of the degradative steps is largely conserved through the coupled production of other chemical bonds. Most important among these bonds is the phosphate bond formed in the conversion of adenosine diphosphate (ADP) to adenosine triphosphate (ATP). This latter compound can donate energy to those steps of biosynthesis requiring energy input, being converted to ADP or AMP (adenosine monophosphate) in the process. Both the particular biosynthetic steps which occur as well as the high rate at which they occur, then, are a consequence of the presence and coordinated action of the cell's enzymes. It is not unreasonable to say that a creature is a reflection of the enzymes which it produces.

Enzymes are protein molecules and as such are composed of one or more identical or different peptide chains, that is, linear arrays of

6 THE MECHANICS OF INHERITANCE

FIG. 1.4a. *The major constituents of a simple growth medium that supports active division of the bacterium* Escherichia coli *at a rate of 1 generation each 40 min at 37°C.*

FIG. 1.4b. *Some of the more notable chemical compounds in a bacterial cell. See William D. McElroy's* Cell Physiology and Biochemistry, *2nd ed. (Englewood Cliffs, N.J.: Prentice-Hall, 1964) for a full discussion of these constituents of cells.*

Enzymes Hundreds of different proteins, each of which is a specific polymer of twenty different **amino acids**

NH₂CH₂COOH
Glycine (Gly)

CH₃
|
NH₂CHCOOH
Alanine (Ala)

CH₃   CH₃
  \\ /
   CH
   |
NH₂CHCOOH
Valine (Val)

CH₃
|
CH₂
|
CHCH₃
|
NH₂CHCOOH
Isoleucine (Ileu)

CH₃   CH₃
  \\ /
   CH
   |
   CH₂
   |
NH₂CHCOOH
Leucine (Leu)

CH₂NH₂
|
CH₂
|
CH₂
|
CH₂
|
NH₂CHCOOH
Lysine (Lys)

CH₂—NH—C—NH₂
| ||
CH₂   NH
|
CH₂
|
NH₂CHCOOH
Arginine (Arg)

HC—N
‖   \\
      CH
C—N
| H
CH₂
|
NH₂CHCOOH
Histidine (His)

    CH₂
CH₂/   \\CH₂
    \\   /
    NHCHCOOH
Proline (Pro)

CH₂OH
|
NH₂CHCOOH
Serine (Ser)

CH₃
|
CHOH
|
NH₂CHCOOH
Threonine (Thr)

COOH
|
CH₂
|
NH₂CHCOOH
Aspartic acid (Asp)

CONH
|
CH₂
|
NH₂CHCOOH
Asparagine (Asn)

COOH
|
CH₂
|
CH₂
|
NH₂CHCOOH
Glutamic acid
(Glu)

CONH₂
|
CH₂
|
CH₂
|
NH₂CHCOOH
Glutamine (Gln)

CH₂SH
|
NH₂CHCOOH
Cysteine (Cys)

SCH₃
|
CH₂
|
CH₂
|
NH₂CHCOOH
Methionine (Met)

OH
|
[benzene ring]
|
CH₂
|
NH₂CHCOOH
Tyrosine (Tyr)

[indole ring]
|
CH₂
|
NH₂CHCOOH
Tryptophan (Try)

[benzene ring]
|
CH₂
|
NH₂CHCOOH
Phenylalanine
(Phe)

Coenzymes Dozens of different kinds of molecules that participate in enyzme-catalyzed reactions, for example,

Nucleic acids Classifiable into two types (DNA and RNA), both of which are polymers containing a five-carbon sugar, phosphorus, and four (or more) different NITROGENOUS BASES

"Structural materials" The cell wall of *E. coli* is composed of several polymers, for example,

and all of its membranes contain lipids, for example,

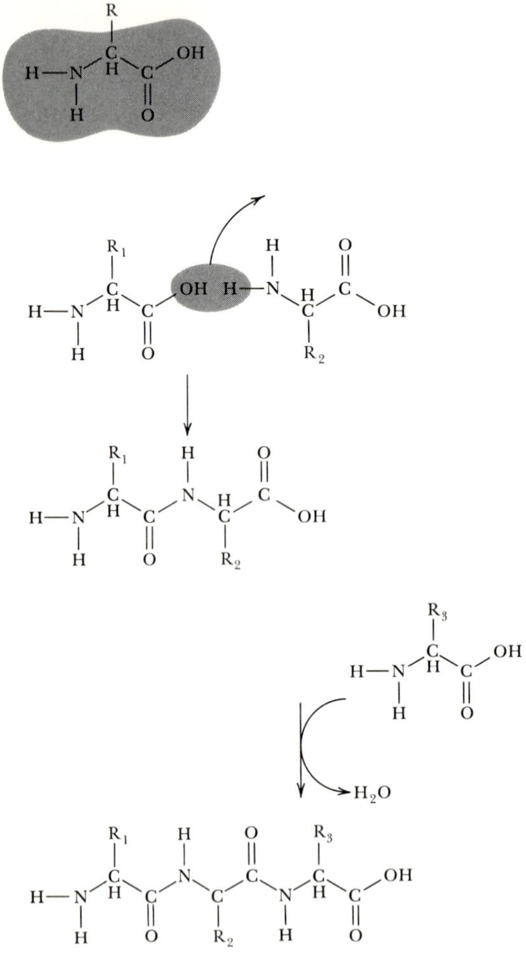

Each of the twenty different amino acids occurring in proteins contains the atomic grouping shown shaded. They differ from each other only in the kind of "R" group attached. (see Fig. 1-4b).

A peptide bond is formed by the elimination of water between two amino acid molecules which condense to form...

a peptide (a *di*peptide in this case).

Addition of another amino acid molecule results in a *tri*peptide. The peptide chains of which proteins are composed are *poly*peptides, the number and sequence of amino acids varying from one kind of protein to another.

The sequential addition of amino acids to peptide chains is itself enzymatically catalyzed in cells. Chapter 10 describes the biosynthesis of proteins in more detail.

FIG. 1.5. **An introduction to polypeptide chemistry.**

amino acids joined to each other by peptide bonds (Fig. 1.5). The catalytic properties of the proteins depend on their shape which in turn is a reflection of the particular configuration into which the peptide chains are folded. The prime factor in determining the chain configuration is the sequence of amino acids in the chain. We shall return later to the question of amino acid sequences in peptide chains (Chap. 10).

Mutation in Bacterial Cultures

Escherichia coli gives rise upon duplication to more *E. coli*. The faithfulness of this reproduction on the morphological level reflects its faithfulness on the enzymatic level. A typical (wild-type) *E. coli* can grow in the simple culture medium shown in Fig. 1.4a, can utilize the sugar lactose as an energy source, and cannot grow in the presence of the antibiotic penicillin. The reproductive fidelity of such typical cells can be challenged by experiment. In such an experiment, a number of wild-type cells are introduced into a culture medium in which all of their descendants can grow even if they lack one of the three qualities characterizing the wild type. The medium contains no penicillin, contains a variety of energy sources, and contains most of the biochemicals of which a *coli* cell is composed. The culture is incubated until millions or billions of cells are present. These cells are then tested for the three properties by which the wild type is characterized.

The cells are easily studied with respect to the reproductive fidelity of the penicillin-sensitivity character. The total number of bacteria in the culture can be estimated by distributing a measured, appropriately diluted volume of the culture on the surface of a nutrient medium solidified with agar. After incubation of the nutrient-agar plate each bacterium has given rise by successive duplications to a clone of bacteria. (A clone is that assemblage of individuals to which a bacterium, or any other creature, has given rise by successive duplications.) The cells in each clone remain clustered together in a colony around the point at which the initial bacterium of the clone was deposited. By counting the colonies, which are easily visible after overnight incubation, the number of cells deposited is determined. From the known dilution and the volume plated, the number of bacteria in the culture can be estimated. The number of penicillin-resistant cells in the culture can be estimated by plating on nutrient agar containing penicillin. We shall define the frequency of penicillin-resistant variants in the culture as (number of resistant bacteria)/(total number of bacteria).

The frequency of variants unable to ferment lactose (lactose-negative variants) can be measured by plating a fraction of the culture on a medium containing lactose and a dye that changes color in the immediate vicinity of colonies that are fermenting lactose.

In each of the procedures above, the frequency of variants in a culture is estimated by counting colonies. Thus the variants that are studied are those which, upon duplication, give rise to variant cells like themselves; they are mutants.

Testing a culture for the frequency of mutants unable to grow on simple culture medium is more difficult. The "replica-plating" method can be applied to this case to ease the labor. About one hundred cells from the culture are plated on a medium of the same properties as the liquid one in which the culture was prepared. The plate is incubated until each cell gives rise to a visible colony. Members of each of the colonies are then transferred to another plate. The transfer is conveniently accomplished, without losing track of the identity of each colony, with the aid of a tautly held piece of velveteen. The velveteen

10 THE MECHANICS OF INHERITANCE

FIG. 1.6. *Replica-plating method for the detection and isolation of mutants unable to grow on simple growth medium. A hundred or so cells from a broth culture of bacteria are spread on a broth medium solidified with agar. After overnight incubation each cell has given rise to a visble colony. A tautly held piece of velveteen is pressed lightly against the plate and then against the surface of a sterile plate containing simple glucose medium solidified with agar. The second plate is incubated until visible colonies appear. The colonies on this replica plate arise in positions*

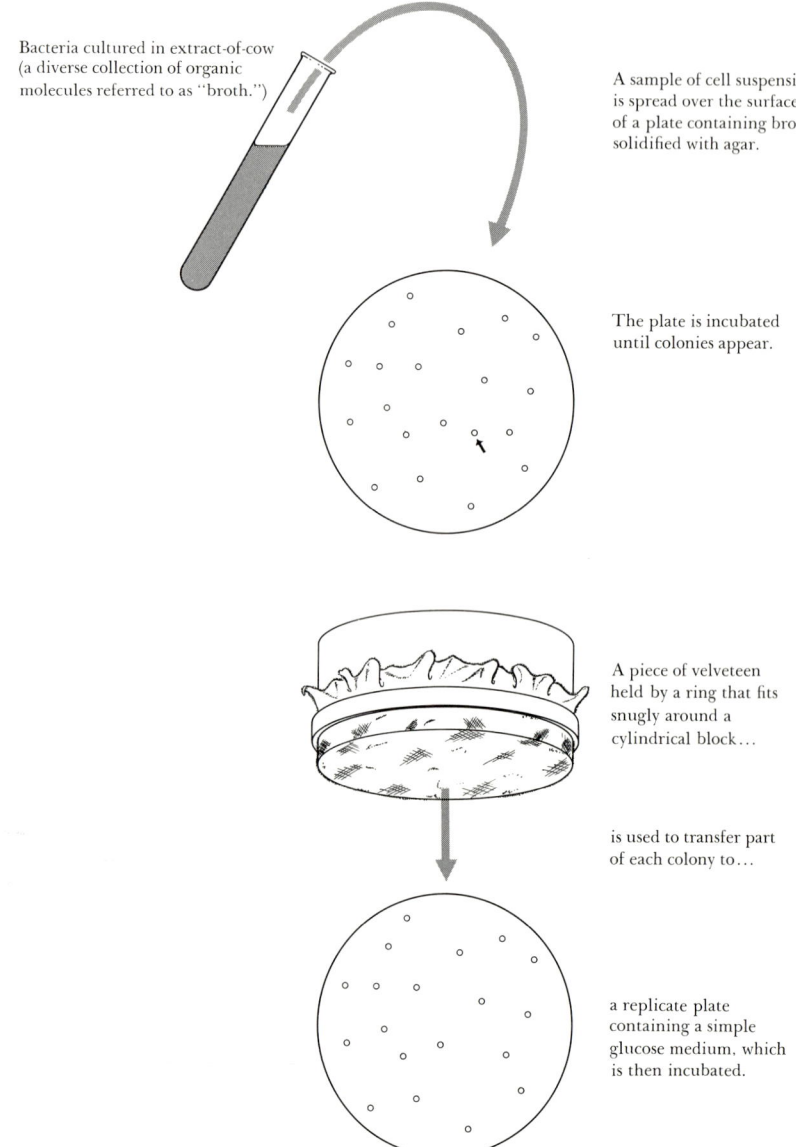

Bacteria cultured in extract-of-cow (a diverse collection of organic molecules referred to as "broth.")

A sample of cell suspension is spread over the surface of a plate containing broth solidified with agar.

The plate is incubated until colonies appear.

A piece of velveteen held by a ring that fits snugly around a cylindrical block...

is used to transfer part of each colony to...

a replicate plate containing a simple glucose medium, which is then incubated.

corresponding to the positions of the colonies on the master plate. Occasional colonies that fail to replicate (such as the one indicated by the small arrow) are nutritionally defective mutants (auxotrophs). The auxotrophs so discovered can be maintained by propagation on broth medium and can be characterized by testing their ability to grow on various defined media. The auxotrophs most useful for further study are those which have simple, easily provided nutritional requirements beyond those of the wild-type cells.

---

is pressed first against the master plate containing the colonies and then against the other (replica) plate. The replica plate, which contains the simple medium described in Fig. 1.4a, is then incubated to permit the formation of visible colonies. The typical outcome of this experiment is that the replica plate has colonies corresponding to each of the colonies on the master plate. Thus each of the hundred cells spread on the master plate has the same simple nutritional requirements as the wild-type cell with which the liquid culture was originally inoculated.

If replication is carried out on a number of master plates, however, the result is not invariably the same. Occasionally a colony fails to replicate, indicating that the growth requirements of its members are not met by the simple medium in the replica plate (Fig. 1.6). An assiduous application of this technique can provide a measure of the frequency of auxotrophic (nutritionally defective) mutants in a culture of prototrophic (nutritionally wild-type) cells.

Measurement of Mutation Rate

The frequency of variant individuals in a population of organisms is determined by many factors. Together with their interactions, these factors are the primary subject matter of *Population Genetics and Evolution,* by Mettler and Gregg, in this series. At this point we are interested in only one of them—the mutation rate.

The mutant frequency in a culture grown from an inoculum of non-mutant cells bears a simple relationship to the mutation rate, the probability per bacterium per division of undergoing mutation.

If we assume the mutation rate of a particular mutant type to be constant per act of duplication, then we can define mutation rate of that type by the equation

$$\Delta m_g = \frac{aN_g}{2} \qquad \text{(Eq. 1.3)}\,[1]$$

where Δm_g is the average number of mutants newly arising in genera-

[1] This and subsequent equations employ a mathematics suitable for synchronously dividing populations. This approach exchanges but a bit of accuracy for a lot of clarity. Three assumptions often justifi-

tion g, a is the mutation rate, and $N_g/2$ is the number of acts of duplication by which the N_g cells in generation g arise from the $N_g/2$ cells of the prior generation. We have to specify Δm_g as the *average* number of mutations occurring in the *g*th generation, because mutations are chance occurrences; there is no guarantee that when $N_g/2$ cells undergo duplication exactly Δm_g mutations will occur. In fact, in a set of identical cultures, the number of mutations occurring in a given generation will show Poisson distribution among those cultures. (Study the Appendix now, if you have not done so before.) The mutation rate a is defined by Eq. 1.3 but is more conveniently determined experimentally by application of either of two expressions which combine Eq. 1.3 with Eq. 1.1, the growth equation.

(1) *Determination of the average number of mutations per culture occurring in a set of cultures started from small inocula.* The population size in each generation is just half the size of the population in the following generation. Thus, the average number of mutations occurring in each generation is just half the average number occurring in the following generation. So, we can write m, the average number of mutations occurring in a culture, as

$$m = \Delta m_g + \frac{\Delta m_g}{2} + \frac{\Delta m_g}{4} + \cdots$$

which for a large number of generations becomes

$$m = 2\Delta m_g = aN_g \qquad \text{(Eq. 1.4)}$$

The average number of mutations per culture in a set of identical cultures started from small (mutant-free) inocula can be determined under the appropriate circumstances. The cultures are grown until an appreciable fraction of the cultures contain some mutant cells, but a comparable fraction is still free of mutants. Since mutations will show Poisson distribution among cultures, we can write

$$P_0 = e^{-m}$$

where P_0 is the fraction of cultures containing no mutant cells. (The graph to be made in Fig. A.2 of the Appendix will give you an instant value of m corresponding to an experimentally measured P_0; or use a slide rule, a log table, or a table of exponential functions.)

(2) *Determination of the average number of mutants per culture in a set of identical cultures.* If we assay a culture at generation g, the

---

able in such experiments are made: (1) Mutant and wild-type bacteria duplicate at equal rates under the conditions employed. (2) The rate of mutation from mutant to wild type is not large compared to the rate of mutation a from wild type to mutant. (3) The total number of mutants for the character under observation is small compared to the total population size.

mutants present will include those that arose in that generation *plus* the descendants of those that arose in previous generations. The number of mutations occurring in generation $g - 1$ is given by the definition of the mutation rate as

$$\Delta m_{g-1} = \frac{aN_{g-1}}{2}$$

However, since

$$N_{g-1} = \frac{N_g}{2}$$

we can write

$$\Delta m_{g-1} = \frac{aN_g}{4} \tag{Eq. 1.4}$$

Each mutant arising in generation $g - 1$ will have duplicated once by generation g so that the number of mutants at generation g which are descended from mutants arising at generation $g - 1$ is

$$\frac{2aN_g}{4} = \frac{aN_g}{2}$$

We see that at generation g, the number of mutants newly arisen (Eq. 1.3) is equal to the number of mutants descended from mutants arising at generation $g - 1$. Convince yourself that we may generalize: On the average, the contribution to the mutant population at generation g from mutations occurring in prior generations is the same for each generation. Thus, on the average, the number of mutants in the culture is

$$\rho = \frac{gaN_g}{2} \tag{1.5a}$$

and the frequency of mutants is

$$\frac{\rho}{N_g} = \frac{ga}{2} \tag{1.5b}$$

The application of Eq. 1.5 is tricky. The equation is valid only when the inocula used to start the cultures from which ρ is to be estimated contain no mutants of the type to be measured. Since large inocula are likely to contain such mutants, small inocula are employed to reduce the chance of including mutant cells among the inocula. However, in a series of identical cultures started from small inocula, mutations will, by chance, occur earlier in some than in others. Mutations occurring quite early give rise to large clones of mutants. Therefore, such early events, although rare, contribute heavily to ρ. In any given set of a few cultures, however, the earliest possible events are

14   THE MECHANICS OF INHERITANCE

unlikely to occur. Therefore, Eq. 1.5a predicts a higher average number of mutants than is likely to be observed. By the same token, the application of Eq. 1.5 to estimates of mutation rate a may lead to an underestimate.[2] Equation 1.5 *is* directly applicable in cases where the cells to be used as inocula have been grown in an environment in which the mutation rate a is small compared to that encountered under the experimental conditions themselves. In this case, one can obtain mutant-free inocula large enough to undergo about the same number of mutations in each of the cultures even in the first generation.

Many factors can influence the mutation rate; some of them are discussed in Chaps. 4 and 10. For now, we simply wish to point out that in ordinary environments mutation rates for various characters are generally found to be low, most of them manifesting reproductive infidelity in far less than 1 per 10,000 duplicative acts.

Mutations and Enzymes

When mutant cells are detected, isolated, and cultivated, their offspring retain the mutant characteristics with a fidelity as great as and often greater than that manifested by the wild type. Mutant cultures can then be examined biochemically to determine what alterations in their enzyme content are responsible for their new properties.

A mutant isolate unable to grow on simple medium may be found to lack the enzyme that catalyzes a known step in the production of a

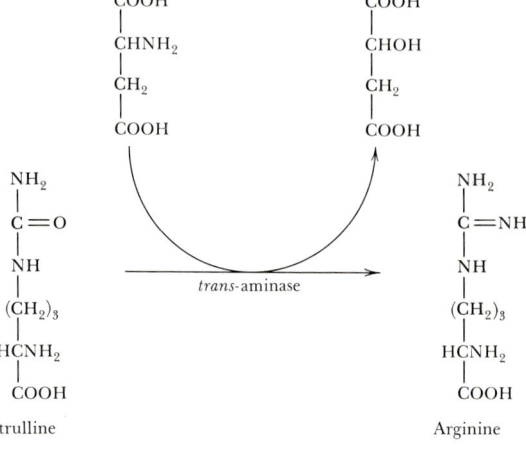

FIG. 1.7. *Mutants that fail to synthesize enzymatically active molecules of the proper trans-aminase cannot make arginine, an amino acid required for protein synthesis.*

[2] Problem 1.6 describes a procedure of Luria and Delbrück for obtaining unbiased estimates of a from experiments using small inocula.

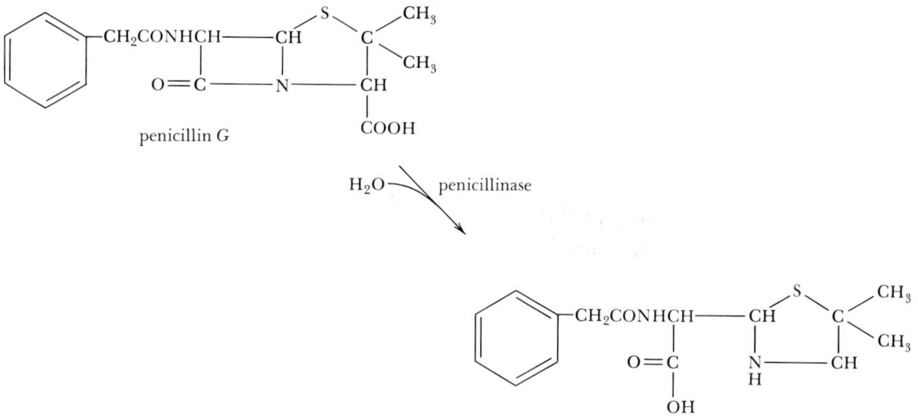

FIG. 1.8. **Strains of bacteria that make large amounts of penicillinase, a protein that enzymatically opens one of the rings on the penicillin molecule, tend to be resistant to the antibiotic.**

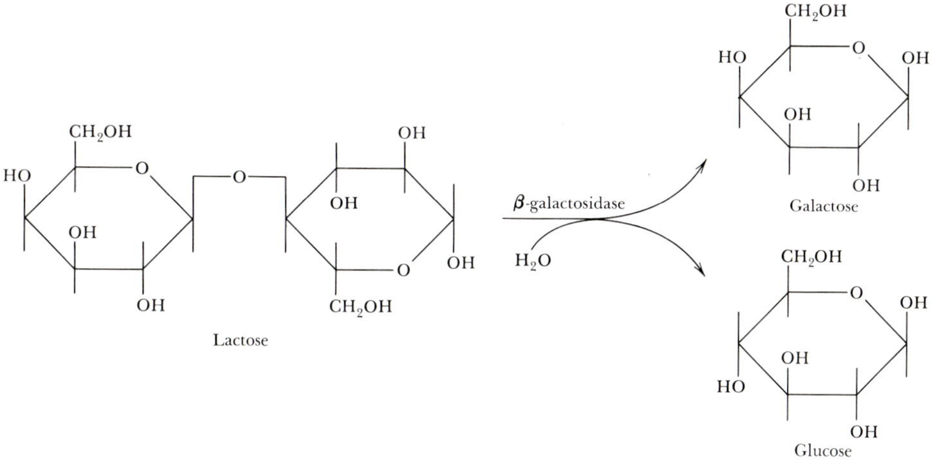

FIG. 1.9. **The disaccharide lactose can be hydrolytically cleaved to galactose and glucose by the enzyme β-galactosidase. Mutants that fail to make this protein, or make an inactive form of it, cannot ferment lactose.**

necessary cellular constituent, for instance, the conversion of citrulline to the amino acid arginine (Fig. 1.7). The penicillin-resistant mutant may be found to contain, in contrast to the wild type, an enzyme that can degrade penicillin to a harmless substance (Fig. 1.8). The mutant strain that cannot ferment lactose may be found to lack the enzyme that hydrolyzes the disaccharide into two monosaccharides—the first degradative step in its utilization as an energy source (Fig. 1.9).

Summary

Three properties of living things are illustrated by the foregoing discussion: (1) The reproduction of living things is faithful, but (2) not perfectly faithful—mutations occur. (3) Many of these mutations alter one of the enzymes of the creature or prevent its appearance. Indeed, only those mutations that *do* change the biochemical capacities of the creature are significant to the creature and detectable to us.

From these considerations a philosopher might build an argument for the existence within living things of a set of instructions with three properties: (1) The instructions are reproduced. (2) The instructions are occasionally altered. (3) The instructions dictate the construction of proteins.

Chapter 2 will examine experiments that identify a substance ("genic material") present within bacteria and other creatures that is responsible for their remarkable reproductive fidelity as well as for their occasional lapses.

References

Lederberg, Joshua, and Esther Marilyn Lederberg, "Replica Plating and Indirect Selection of Bacterial Mutants," *J. Bacteriol.*, 63 (1952), 399–406. Reprinted in *Papers on Bacterial Genetics*, Edward A. Adelberg, ed. (Boston: Little, Brown & Co., 1960), pp. 24–31. A description of the technique and several applications of the replica plating method.

Luria, S. E., and Max Delbrück, "Mutations of Bacteria from Virus Sensitivity to Virus Resistance," *Genetics, 28* (1943), 491–511. Reprinted in *Papers on Bacterial Genetics*, pp. 3–23. The mathematical basis for the analysis of mutation in bacterial cultures.

McElroy, William D., *Cell Physiology and Biochemistry*, 2nd ed. Englewood Cliffs, N.J.: Prentice-Hall, Inc., 1964. An efficient review of cellular biochemistry at the beginning level.

Problems

Answers for problems begin on page 207.

1.1. The graph (next page) depicts the increase in the number of cells in a bacterial culture after inoculation of nutrient-broth medium with 1.5×10^3 cells from an "old" culture. After a lag of about a half hour, the culture becomes actively dividing. For the portion of the curve that is exponential, we define the generation time as the time within which the population exactly doubles in size. Calculate the generation time for the growth curve depicted.

1.2. Suppose we inoculate 100 ml of broth with 1,000 actively dividing cells each of bacterial strain A and strain B. A few hours later we observe 5×10^5 per ml of strain A and 5×10^3 per ml of strain B. By what factor is the growth rate of strain A greater than that of strain B?

Heredity

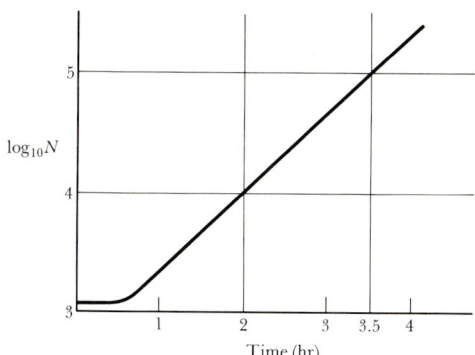

1.3. Liquid bacterial growth medium in a sterile tube was inoculated with actively multiplying bacteria. Call the time of inoculation time-zero. The culture was assayed for the number of bacteria each hour thereafter. The data are tabulated below.

Time	Cells
0	3.0×10^5
1	8.0×10^5
2	2.0×10^6
3	5.3×10^6
4	1.4×10^7
5	3.6×10^7
6	9.4×10^7
7	2.5×10^8

(a) On a sheet of graph paper plot log (number of bacteria) versus time.
(b) What is the time required for the number of cells to double?
(c) Suppose there were *no* mutants (of a specified type) in the inoculum. When the number of bacteria reached 1.2×10^6, the frequency of mutants was measured and found to be 10^{-4}. How many mutant cells would you expect to find when the number of cells reached 2.4×10^6? (Assume a negligible rate of back mutation and equal viability of the mutant and wild type.)
(d) What is the value of the mutation rate?

1.4. Each of 10^6 tubes was inoculated with 1 penicillin-sensitive cell. Growth then proceeded in perfect synchrony. When there were 1,024 cells in each tube, the tubes were examined for penicillin-resistant cells. One thousand of the tubes had at least 1 resistant cell. On the assumption that penicillin resistance arises by mutation, how many tubes would you expect to find with *exactly* 1 resistant cell? Two resistant cells? Four resistant cells? More than 4 resistant cells? (In this, as in every problem, look first to your own understanding. The first thing you should spot here is that in practically none of the tubes did two mutations occur. Now you're on your own.)

1.5. Each of 100 identical tubes of growth medium was inoculated with about 10^4 streptomycin-sensitive bacterial cells. When the cultures reached 10^8, they were examined for the presence of streptomycin-resistant cells; 87 of the tubes contained at least 1 such cell.
(a) What was the average number of mutations per tube?
(b) Calculate the mutation rate to streptomycin resistance.

1.6. S. E. Luria and M. Delbrück (see References) modified Eq. 1.5 such that it would give unbiased estimates of mutation rates when applied to data from a set of cultures started from small, mutant-free inocula. In an experiment involving only a few cultures, mutations in the early generations are unlikely to occur. One would usually be better off, then, to assume that the early generations do not make their contribution to the final mutant population. Therefore, rather than calculating a from Eq. 1.5, it is better to use an equation which describes the "likely" average number of mutants.

If C cultures are employed in the experiment, it is reasonable to start counting generations when, on the average, there have been 0.5 mutations somewhere in the set of cultures, that is, when

$$Cm = CaN = 0.5$$

where N is the population size in each culture when $Cm = 0.5$. The growth equation permits us to write N in terms of the final population size N_g and the number of generations (G) required to increase the population size from N to N_g

$$N = \frac{N_g}{2^G}$$

so that

$$Cm = 0.5 = \frac{CaN_g}{2^G}$$

which gives us

$$G = \log_2 2CaN_g$$

Substituting G for g in Eq. 1.5a gives us the "likely" average number of mutants (r) in a set of C cultures

$$r = \frac{aN_g}{2} \log_2 2CaN_g$$

Now, in a set of 8 cultures grown from small mutant-free inocula, suppose Luria and Delbrück observed an average of 8 mutants per culture. The final population size in each culture was 10^8. What was their estimate of the mutation rate? (*Hint:* tabulate r for selected values of aN_g. Graph r versus aN_g over a range which includes $r = 8$. From the graph read off the value of aN_g corresponding to $r = 8$.)

Two

The Genic Material

The series of experiments that led to the identification of the genic material can be said to have begun about 1920. It was about that time that both bacterial viruses and bacterial transformation were discovered.

Bacterial Transformation

Wild-type cells of the pneumococcus bacterium (*Diplococcus pneumoniae*) are encapsulated by a polysaccharide excretion. If such cells are introduced into a susceptible mammal, severe illness usually follows. Occasionally mutants devoid of a capsule arise in laboratory cultures of pneumococci. Such mutants are less virulent than the encapsulated strains; they rarely cause an animal any difficulty. The study of encapsulated and nonencapsulated pneumococci led F. Griffith to the discovery in the 1920s of bacterial transformation. Griffith's experiment demonstrating transformation of avirulent bacteria to the virulent type is summarized diagramatically in Fig. 2.1.

Griffith inoculated one group of mice with avirulent cells, a second group with heat-killed virulent cells, and a third group with both avirulent and heat-killed virulent cells. The first two groups of animals were unaffected by the inoculations. Mice in the last group, however,

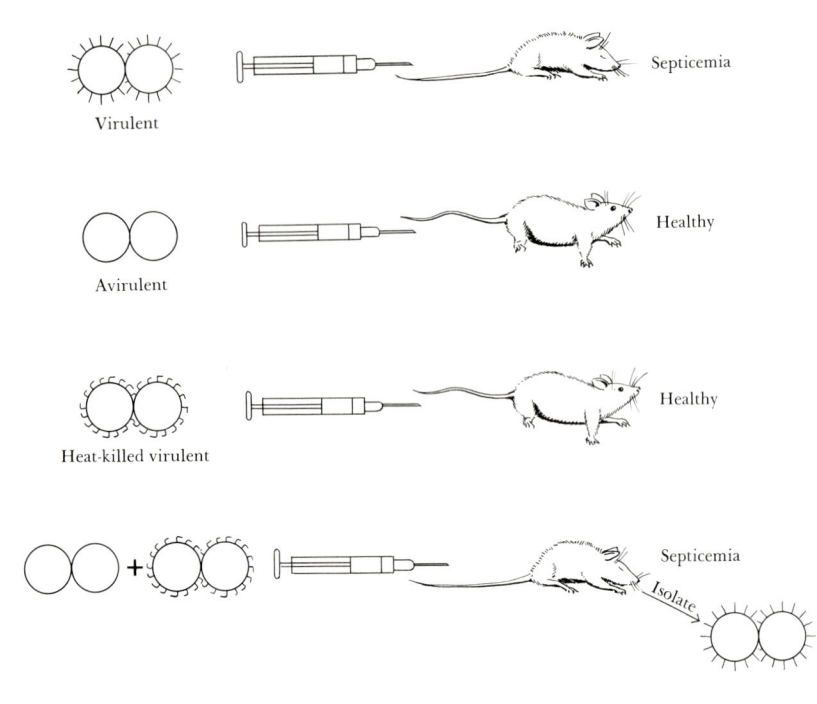

FIG. 2.1. *Bacterial transformation in living animals (in vivo). Avirulent cells and heat-killed virulent cells interact in the mouse to produce living virulent cells. This figure is based on Fig. 5 in David M. Bonner's* Heredity *(Englewood Cliffs, N.J.: Prentice-Hall, 1961).*

developed septicemia. The interaction within the mice between viable avirulent cells and heat-killed virulent cells led to the appearance of viable virulent cells. Two alternative explanations for this result were prominent. (1) The presence of viable avirulent pneumococci restored the heat-killed cells to viability. (2) The heat-killed cells conferred the property of virulence upon the avirulent cells. The latter conclusion is the proper one, but its proof follows most easily from experiments performed subsequently and described below.

A major advance in the study of bacterial transformation came with the demonstration in the early 1930s that transformation can be obtained in vitro. When a nonencapsulated cell population was exposed to the debris from disrupted encapsulated cells, a small fraction of the cells were transformed to the encapsulated virulent type.

In recent years, transformation of a number and variety of hereditary characters has been demonstrated among different species of bacteria. In many of these cases the character transformed represents the ability to make a protein that functions as an enzyme in the catalysis of an identifiable chemical conversion within the cell. For the case of encapsulation, the presence or absence of a capsule depends directly on the presence or absence of a functioning enzyme required in one of the steps in capsule formation. In the bacterium *Bacillus subtilis* it has been possible to transform each of the types of mutant characters illustrated in Figs. 1.6, 1.7, and 1.8. An in vitro transformation experiment using *B. subtilis* is diagrammed in Fig. 2.2.

Bacterial transformation manifests the following three properties: (1) In order to transform cells of, say, type X to type Y (where X and Y are mutually exclusive alternatives), debris from disrupted cells of type Y is required. Cells of type X do not yield debris capable of inducing transformation from X to Y. (2) When the type Y cells derived by transformation multiply, they give rise to more type Y cells. When disrupted, the progeny of transformed cells yield debris that can transform X cells to Y cells; this debris will *not* transform Y cells to X cells. (3) Transformation can be carried out reciprocally. If debris from type Y cells can transform type X cells to type Y, then debris from type X cells can transform type Y cells to type X.

Each of these three points will be subject to amplification in subsequent chapters. At this point we need to establish only that *transformation involves a specific, directed alteration in a hereditary characteristic of a bacterium.* An elucidation of the mechanism of transformation was certain to have a profound impact on our views of heredity.

FIG. 2.2. *Bacterial transformation in the test tube (in vitro). Debris from cells* (**Bacillus subtilis**) *capable of growing on a simple medium can transform nutritionally defective (auxotrophic) mutants to prototrophy (state of being hereditarily wild type with respect to nutritional requirements).*

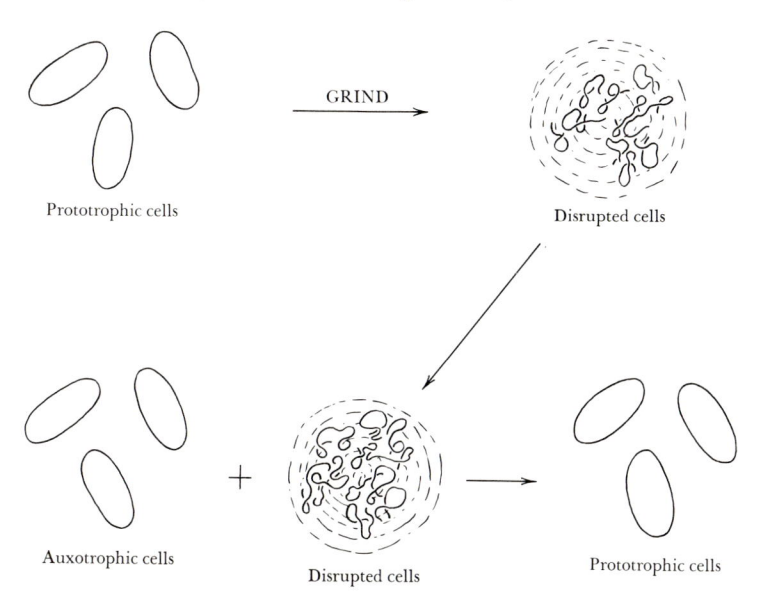

22 THE MECHANICS OF INHERITANCE

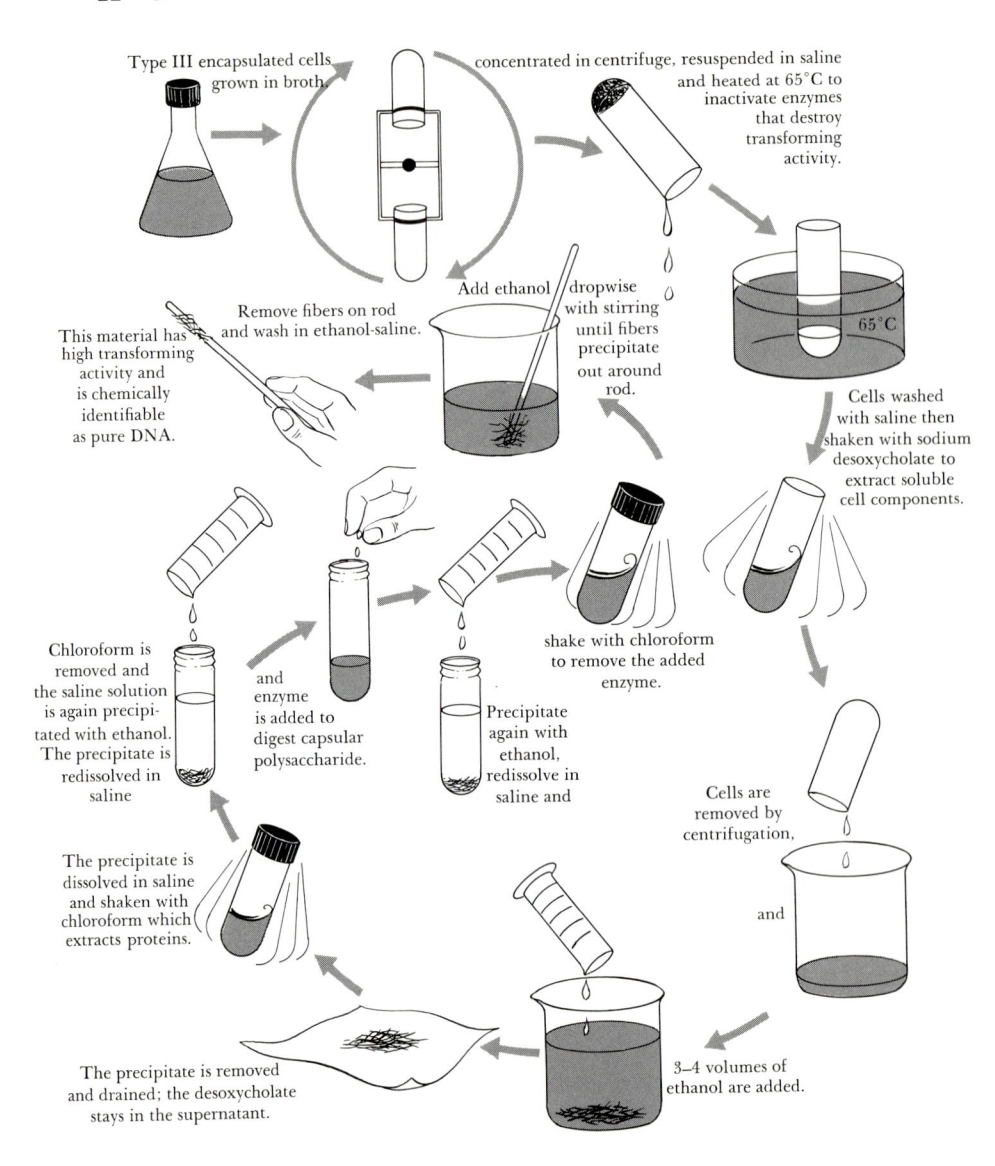

FIG. 2.3. *The procedure by which Avery, MacLeod, and McCarty isolated pure DNA with high transforming activity. The DNA derived from cells encapsulated in type III polysaccharide was added to cells of a nonencapsulated strain derived by mutation from type II cells. Some of these cells were transformed by the DNA to the type III encapsulated variety.*

Identification of the Transforming Agent

In 1944, O. T. Avery, C. M. MacLeod, and M. McCarty announced the results of ten years of work in the analysis of the mechanism of transformation. They fractionated into various chemical species the debris from disrupted encapsulated cells and determined the transforming activity of each. The procedure and outcome of their experiment are outlined in Fig. 2.3.

Only the DNA fraction (see Fig. 1.4b) was found capable of transforming nonencapsulated cells to encapsulated cells. The experiment has by now been repeated a multitude of times for the character of encapsulation as well as for many of the other hereditary characteristics that occur in systems amenable to transformation.

Two principal hypotheses were advanced to account for the role of DNA in transformation. One hypothesis supposed that DNA could act somehow upon genic material to produce directed mutations, that is, to change the genic material of the recipient cells to make it resemble the genic material of the donor cells. The second hypothesis *equated* DNA with the genic material and explained transformation as a transfer of genic material from donor to recipient cells.

The identification of DNA as genic material was further substantiated in an experiment performed on viruses that multiply inside of bacteria (bacteriophage).

The Life Cycle of Bacteriophage T2

Bacteriophages (phages), like other viruses, multiply inside of cells giving rise to more phage particles with a fidelity comparable to that of the cat or the *coli*. About twenty minutes after a bacterium is infected by a phage particle, it bursts open, liberating a few hundred particles identical (usually) to the particle that initiated the infection. The phage T2, which multiplies inside *E. coli*, was used in the experiment that aided in the identification of DNA as genic material.

Particles of T2 phage are composed of about 50 percent by mass of protein and 50 percent DNA. In 1952, A. D. Hershey and Martha C. Chase demonstrated that when a phage infects a bacterium, its DNA enters the host cell, but over 90 percent of its protein remains attached to the outside surface of the cell. Most of the protein on the outside of the cells can be removed by shear forces vigorously applied throughout a suspension of infected bacteria. (The technical details of this experiment are described in the legend to Fig. 2.4.) Infected bacteria from which the protein moiety of the T2-phage particles had been removed produced a normal crop of T2-phage particles; *the DNA that entered the cell constituted the sole physical thread of hereditary continuity between infecting and emerging particles.*

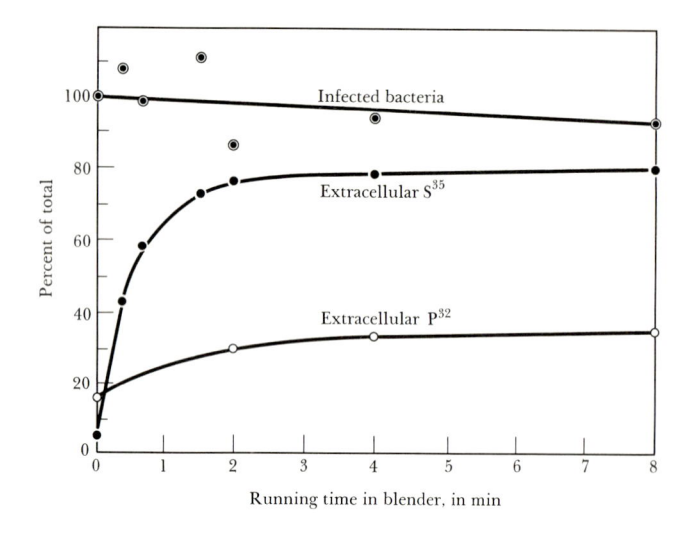

FIG. 2.4. *Two batches of phage particles were prepared, one batch by growing phages in bacteria containing radioactive sulfur (S^{35}), the other by growing bacteria containing radioactive phosphorus (P^{32}). Since several amino acids contain sulfur but DNA does not, the first batch of phage particles was labeled in its protein moiety; since phosphorus is a component of DNA (see Chapt. 3) but not of proteins, the second batch was labeled in its DNA. Each batch of phages was permitted to attack bacteria. The infected bacteria were subjected to shear forces in a kitchen blender, and the amount of radioactivity removed from the cells was measured. The S^{35} was easily removed, but the P^{32} was not. The number of infected bacteria that produced a crop of phage particles was diminished by the treatment only slightly, if at all. This figure, reproduced by permission of A. D. Hershey, originally appeared in J. Gen. Physiol., 36 (1952), 47.*

The Hershey–Chase experiment, when considered together with the observations of many other workers, leads to the picture of a phage infection outlined in Fig. 2.5. The phage particles attach to the bacterium by means of the "tail." The DNA, contained within the head made of protein, enters the cell. Within twenty minutes about two hundred new phage particles are formed complete with DNA and protein "coat." Each of these particles can itself initiate a full cycle of infection if permitted to attack a bacterium.

In organisms other than bacteria and viruses the evidence regarding

FIG. 2.5. *The infectious cycle of a virulent bacteriophage as illustrated by phage T2. The mature particle, shown in longitudinal section, attaches by its tail fibers to a bacterium. The sheath contracts, the core penetrates the cell surface, and the DNA of the particle passes into the bacterium. About 20 min later at 37° C the cell bursts and a few hundred new mature particles are released.*

The Genic Material 25

26 THE MECHANICS OF INHERITANCE

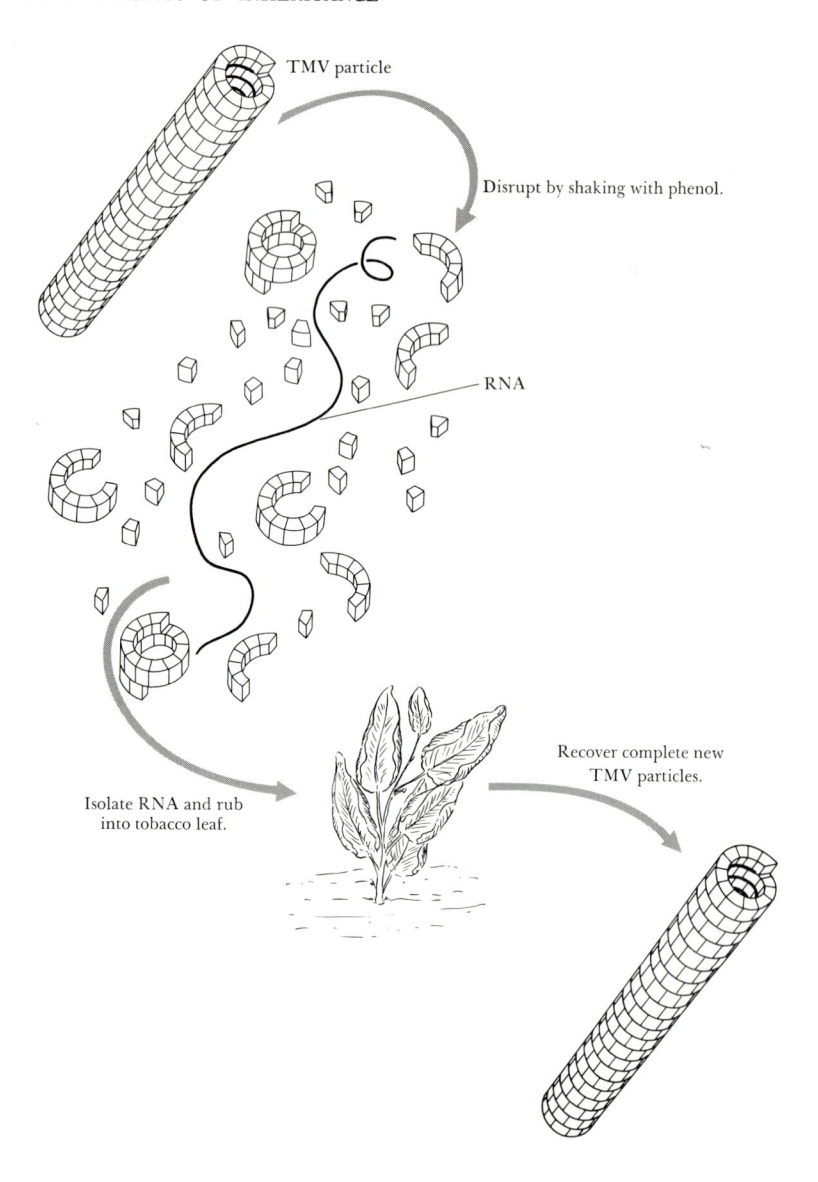

FIG. 2.6. *Tobacco mosaic virus (TMV) multiplies inside cells of plants in the tobacco family. The virus particles are rod-shaped and are composed of a protein sheath surrounding an RNA core. The protein can be dissociated from the RNA by shaking the particles with water and phenol. The RNA and protein can then be separately tested for their ability to infect tobacco plants. Only the RNA fraction of the virus retains this ability. The plants infected by this RNA produce a crop of complete TMV particles.*

the DNA nature of the genic material is primarily circumstantial. For instance, the sperm of multicellular animals, a cell specialized as a motile package of genic material, is relatively rich in DNA. Other evidence relating to the nature of the genic material of multicellular organisms is implied in subsequent chapters. Throughout this book it will be our working hypothesis that DNA is genic material wherever it appears and that every creature's genic material is DNA until proven otherwise. (For a number of viruses it *has* been proven otherwise.)

RNA Viruses

Some of the smallest viruses contain no DNA, but instead are made of RNA in addition to protein. *Gene Action* by Hartman and Suskind, another volume in this series, has a lot to say about RNA; subsequent chapters of this volume have things to say about it too. However, at this point we need know only that RNA is a close chemical relative of DNA; both are nucleic acids. For a number of the RNA viruses, as well as for the smaller DNA viruses, successful infections have been initiated by application to the host organisms of purified nucleic acid isolated from the particles. The first such experiment was performed with tobacco mosaic virus (TMV); it is diagrammed in Fig. 2.6.

The Structure of DNA

*D*eoxyribo*n*ucleic *a*cid (DNA) usually occurs in nature as long unbranched, polymeric molecules (Fig. 2.7). A DNA molecule is composed of an intimately associated pair of DNA chains. Each DNA chain commonly is composed of four kinds of monomers.

The monomers out of which a DNA chain is composed are called nucleotides or, more specifically, deoxyribonucleotides. Each nucleotide is a compound of phosphoric acid, the 5-carbon sugar deoxyribose, and one or another of the four kinds of nitrogenous bases. The phosphoric acid and deoxyribose are combined by an ester linkage. For our present purposes we may consider the linkage to be at the number 5 carbon atom of the deoxyribose (Fig. 2.8). The nitrogenous bases, whose structural formulas are shown and identified in Fig. 2.9, are each linked to a deoxyribose at the number 1 carbon atom of the sugar. The structural formulas of the various nucleotides are shown in Fig. 2.10.

In a DNA chain, the nucleotides are linked to each other by ester bonds between the phosphate group of each nucleotide and the number 3 carbon in the deoxyribose of the adjacent nucleotide. A short stretch of a DNA chain showing five nucleotides selected willy-nilly is diagrammed in Fig. 2.11. A DNA chain can be thought of as having a sugar-phosphate "backbone" (on the left in Fig. 2.11) to which is attached at each level one or another of the nitrogenous bases. In nature, DNA chains have various lengths. Some chains are thought to

28

FIG. 2.7. *An electron micrograph of part of a DNA molecule. A DNA molecule is 20 Å wide and extremely long. This photograph was kindly supplied by Michael Beer. Magnification 100,000×.*

FIG. 2.8. **Deoxyribose phosphate,** *polymers of which form the "backbone" of a DNA chain.*

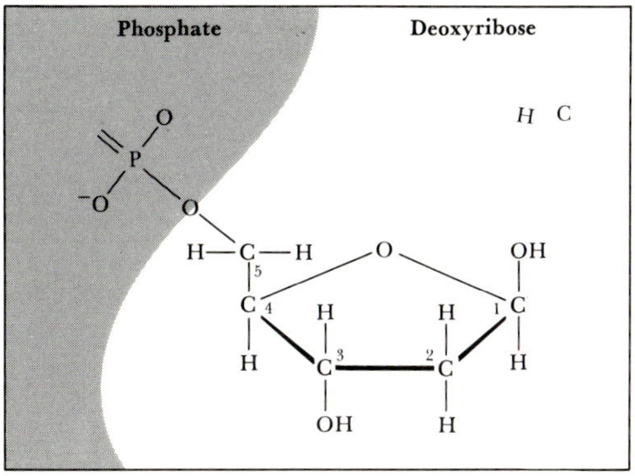

FIG. 2.9. *The four nitrogenous bases most commonly found in DNA. The two pyrimidines have a common heterocyclic ring "nucleus," as do the two purines.*

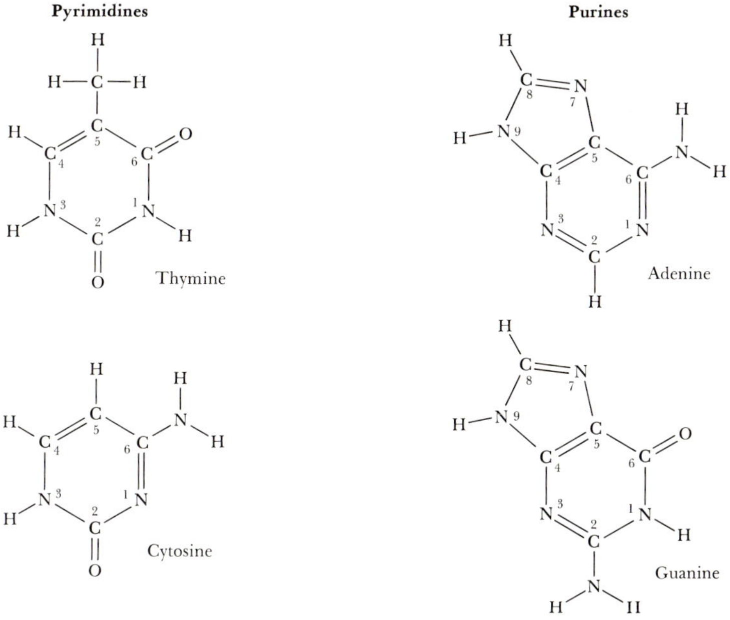

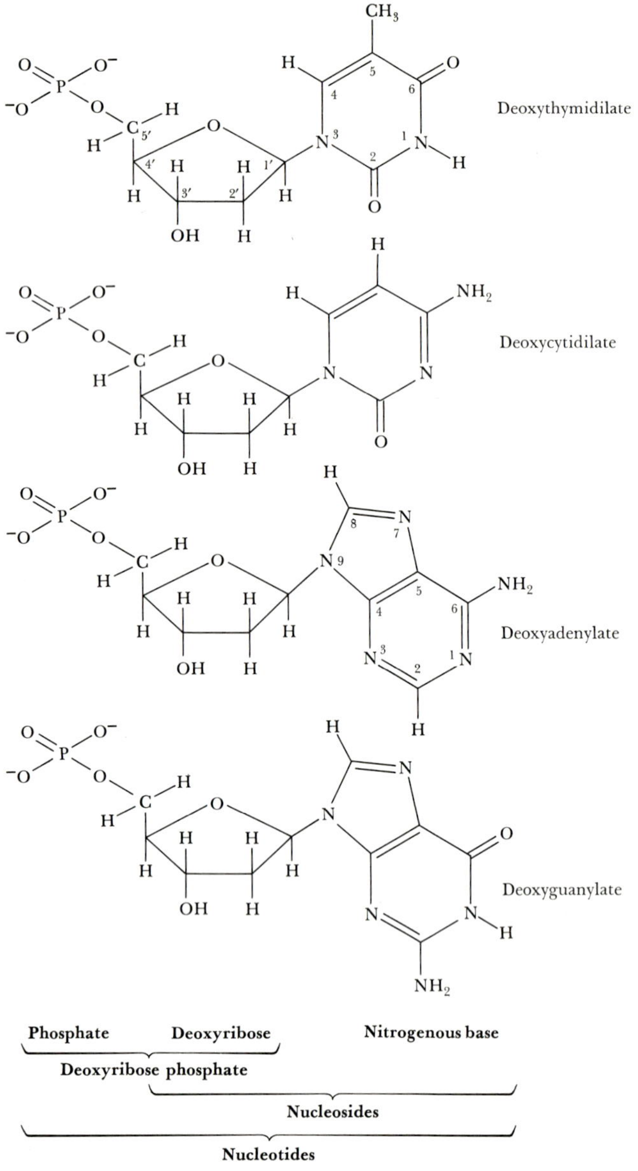

FIG. 2.10. *The four common deoxyribonucleotides. The nucleotides of DNA are composed of the nitrogenous bases each attached to deoxyribose phosphate. A nucleotide devoid of its phosphate is called a nucleoside. Each of the nitrogeneous bases is attached to deoxyribose by a bond that connects a ring-nitrogen atom of the base to the number 1 atom of the deoxyribose. This latter ring position is often referred to as the number 1' position of the nucleoside. Henceforth we shall use one of the conventional symbolisms of the organic chemist: carbon atoms in ring structures are implied by a bond angle.*

FIG. 2.11. *A stretch of DNA with a nucleotide sequence chosen willy-nilly. The nucleosides are connected by phosphate "bridges" that run from the 5' position of one nucleoside to the 3' position of the next. In this figure, and henceforth, we adopt another convention of the organic chemist: H's are left out of the structural formulas unless there is a reason for drawing attention to them.*

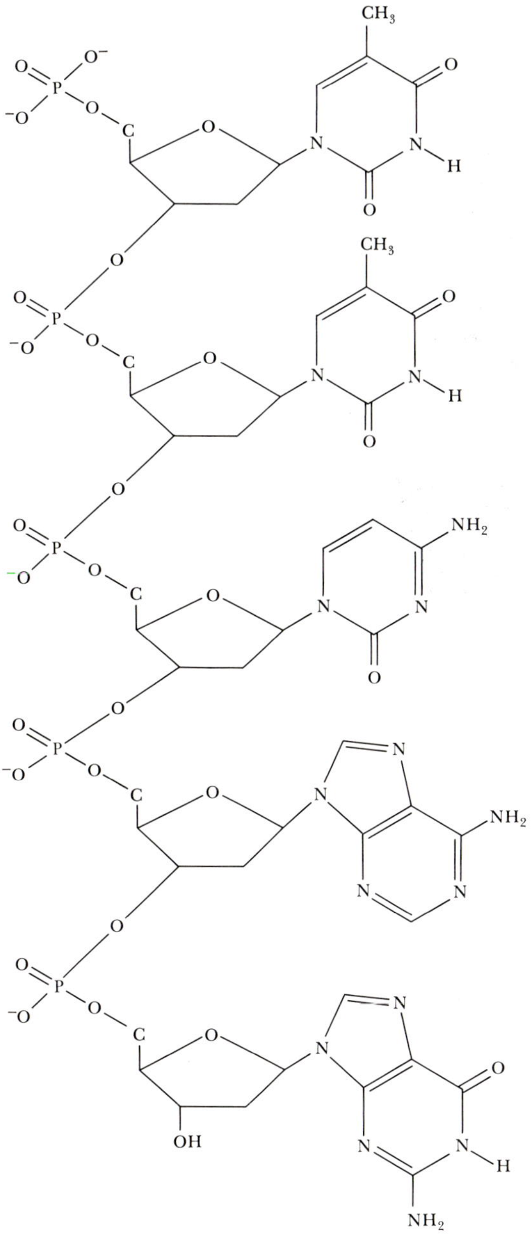

31

32 THE MECHANICS OF INHERITANCE

be composed of 200,000 nucleotides; longer ones probably exist (see Chap. 5).

DNA molecules consist of two chains twisted about each other. There are about ten nucleotides in each chain for each complete turn of the double helix. Each nucleotide in a chain is oriented with its nitrogenous base toward the other chain and its phosphate group away from the other chain. The two chains are held together by hydrogen bonds formed between bases that occupy the same level on the two chains.

A hydrogen bond results from the sharing by two atoms of a single hydrogen atom. Compared to ordinary covalent bonds, hydrogen bonds are very weak. In Chap. 3 we shall see that the two chains of a duplex have occasion to separate from each other. The separation is obviously facilitated by the weakness of the hydrogen bonds.

In a hydrogen bond, one participating atom acts as a hydrogen donor and the other as a hydrogen acceptor. Each of the nitrogenous bases contains donor atoms and acceptor atoms. Thus, a great variety of hydrogen-bonding arrangements between pairs of bases is possible. In a normal DNA molecule, however, there are restrictions imposed upon the hydrogen-bond interactions between bases. These restrictions are dictated to some extent by the rules of formation of the covalent bonds that join the bases to the deoxyribose and that unite the nucleotides to each other along the sugar-phosphate backbone. Additional restrictions are imposed by three observations. (1) All the base pairs are hydrogen bonded. (2) The sugar-phosphate backbones form a very

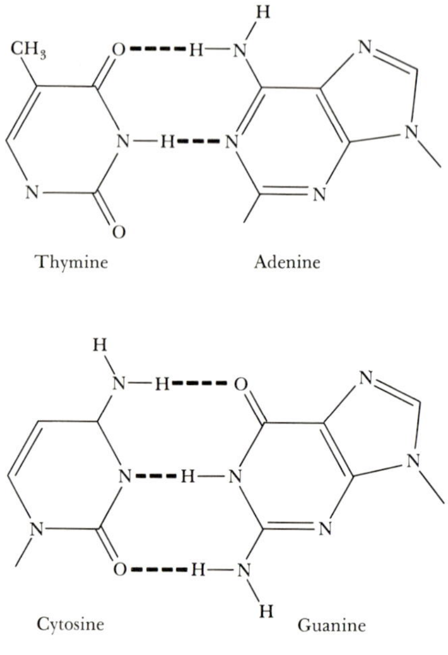

FIG. 2.12. *Hydrogen bonding between bases as it occurs in DNA. Of the various hydrogen-bonding interactions that can occur among bases, only these two normally occur in DNA.*

regular double helix. (3) The planes of the bases are perpendicular to the long axis of the molecule. These three considerations limit to two the number of base-pair associations that can (normally) occur in a DNA molecule. These are shown in Fig. 2.12.

A diagrammatic representation of the structural formula for a short stretch of a DNA molecule is shown in Fig. 2.13. The same base sequence used in Fig. 2.11 is shown with the half twist in the stretch of molecule "ironed out" to simplify the representation. A diagrammatic representation of a stretch of DNA, complete with twist, is shown in Fig. 2.14.

The solution of the structure of DNA was made possible by the work of hosts of chemists and physicists. The data that permitted the final assembly of these facts were obtained by M. Wilkins and his collaborators and by R. E. Franklin and R. G. Gosling. They used the technique of X-ray diffraction, which can, with rather severe limitations, give indications of the spatial arrangements of some of the atoms within a large molecule. Because of the limitations in the technique, X-ray diffraction data of DNA needed lots of interpreting; J. D. Watson and F. H. C. Crick performed that job. The essential accuracy of their interpretation has been attested to repeatedly by the acquisition of more and better diffraction data. Other observations, described below, have strengthened our belief in the validity of the Watson–Crick structure for DNA.

A feature of the Watson–Crick model for DNA is that at any level where adenine appears on one chain, thymine appears on the other, and wherever guanine appears on one chain, cytosine appears on the other. The obvious implication of this feature is that for DNA molecules isolated from any source whatever, the molar amount of adenine is equal to that for thymine ($A = T$), and the molar amount of guanine is equal to that for cytosine ($G = C$). There is no feature of the model that restricts the relative amounts of AT pairs to GC pairs. In fact, such variation in relative amounts of A and G (or T and C) is a commonplace observation from chemical analysis of DNA. These same analyses of base composition of DNA provide the chemical evidence in favor of the postulated equalities, $A = T$ and $G = C$. A bit later we shall encounter some small viruses for which these equalities do not hold. These virus particles turn out to contain single-chain DNA.

An experiment from the laboratory of A. Kornberg adds another type of confirmation to the Watson–Crick model. Enzymes occur in nature that can hydrolytically cleave the phosphate-sugar ester bonds of DNA backbones. One of these enzymes cleaves only the bond connecting the number 5 carbon of each sugar to phosphate. If hydrolysis is permitted to proceed to completion, the DNA molecules in a treated solution are all cleaved to nucleotides. (Hydrogen bonds between bases are not strong enough to keep the two members of a single bonded pair together throughout the steps of the procedure.) These nucleotides differ from those shown in Fig. 2.10 in that each phosphate group is

34 THE MECHANICS OF INHERITANCE

FIG. 2.13. *A stretch of a DNA molecule. The chain on the left has the same willy-nilly base sequence shown in Fig. 3.5. The chain on the right has a base sequence complementary to it. The two chains, whose "backbones" run in opposite directions, are held together by hydrogen bonds between the bases. The molecule is further stabilized by its helical nature; the chains are twisted about each other with one twist for each ten nucleotide pairs.*

FIG. 2.14. *A schematic representation of a DNA molecule. The relationship of the two hydrogen-bonded chains to each other and to the helix axis is illustrated and the basic physical parameters of the model are given.*

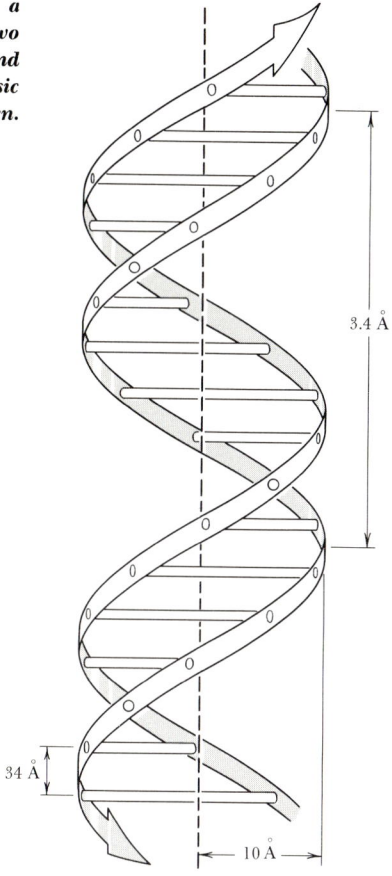

attached to the number 3 carbon of the sugar rather than to the number 5 carbon. This enzymatic tool makes it possible to determine the frequency with which the four nucleotides occur as nearest neighbors to each other along the chains of DNA molecules. The experimental analysis follows.

Adenine nucleotide (for instance) labeled with radioactive phosphorus (P^{32}) in the phosphate group attached to the number 5 carbon of the sugar is "fed" to a system (see Chap. 3) that is synthesizing **DNA**.

DNA is then isolated from the system and enzymatically cleaved at the number 5 carbon atoms. The nucleotides are next isolated and separated into the four kinds. The amount of radioactivity in each type of nucleotide is determined. The radioactivity shown by a given type of nucleotide must be proportional to the frequency with which that nucleotide occurs bonded by the phosphate group running from its sugar-carbon atom 3 to the 5 carbon in the sugar of an adenine nucleotide. The experiment is repeated by feeding each of the other radioactive nucleotides, hydrolyzing, and measuring radioactivities of the isolated nucleotides. The data considered collectively provide an estimate of the frequency with which each of the nearest-neighbor associations shown here occur ((P) is phosphate). You may deduce other relations for yourself. The outcome of the experiments was in accord with the predicted relationships.

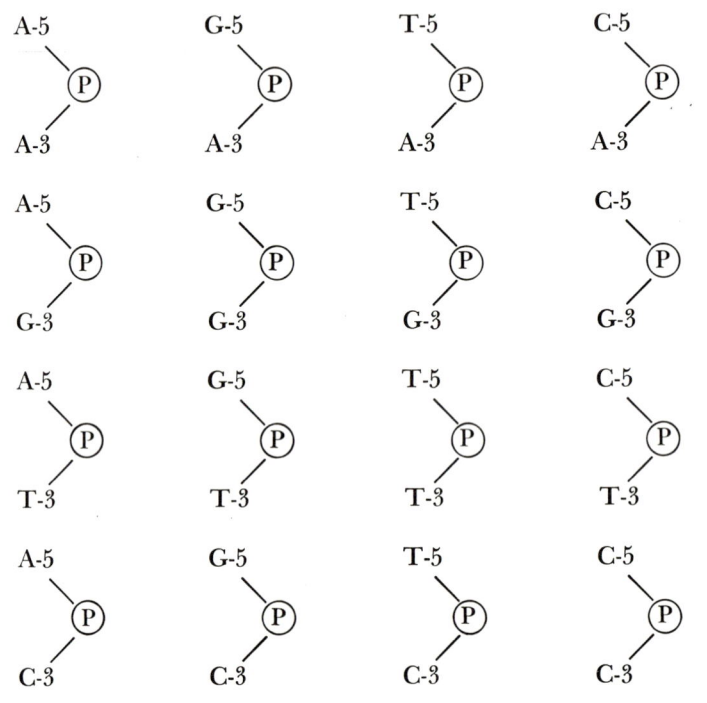

A-5\(P)/A-3	G-5\(P)/A-3	T-5\(P)/A-3	C-5\(P)/A-3
A-5\(P)/G-3	G-5\(P)/G-3	T-5\(P)/G-3	C-5\(P)/G-3
A-5\(P)/T-3	G-5\(P)/T-3	T-5\(P)/T-3	C-5\(P)/T-3
A-5\(P)/C-3	G-5\(P)/C-3	T-5\(P)/C-3	C-5\(P)/C-3

To understand the outcome of these experiments we must refer to the diagram of a DNA chain in Fig. 2.11. Note that a DNA chain has polarity; on one end of the chain the terminal nucleotide is attached to the chain at the number 3 carbon atom of its sugar, while the terminal nucleotide at the other end is attached at its number 5 carbon atom. If we look now at Fig. 2.13, we see that in a DNA molecule, the two chains run in opposite directions. This feature of the model makes special predictions about the relative frequencies of nearest-neighbor associations. Some of the relations to be expected are evident from an examination of Fig. 2.13; for example,

```
A-5        T-3                    G-5        C-3
  \         \                       \         \
   P  =      P                       P  =      P
  /         /                       /         /
A-3        T-5                    T-3        A-5


A-5        T-3                    T-5        A-3
  \         \                       \         \
   P  =      P          .            P  =      P
  /         /                       /         /
G-3        C-5                    C-3        G-5
```

You may deduce other relations for yourself. The outcome of the experiments was in accord with the predicted relationships.

The opening paragraph of this chapter qualified the statement that DNA occurs in nature as a pair of chains. Some very small bacteriophages contain only single DNA chains. After the DNA from these phages invades a bacterial cell, however, it becomes double-chained like other DNA. This behavior points up one feature of DNA implied previously in this chapter. Since DNA is genic material, it must contain in some fashion the information for ordering the amino acids in proteins and otherwise directing the life cycle of each cell. As pointed out by Watson and Crick, the only reasonable source for such specificity is in the sequence in which the nucleotides occur within a DNA molecule. (We shall see ample verification for this surmise in Chap. 4 and in later chapters.) The sequence of nucleotides on one chain of a DNA molecule has point-to-point complementarity with the sequence on the other chain. The two chains, then, carry exactly the same information. The little phages containing single DNA chains seem merely to have reduced the genic baggage that they must carry in their quest for a host cell.

The Rare Deoxyribonucleotides

The opening paragraph of this chapter hedged also with respect to the nucleotide composition of DNA. Although cytosine, thymine, adenine, and guanine are the four most common bases, several others do occur in various DNA's. Each of these rarer bases substitutes for one of the common bases. Thus, for instance, phage T2 contains no cytosine but instead contains 5-hydroxymethyl cytosine. The amount of this latter base in T2 is equal to the amount of guanine. It is apparent, therefore, that in T2 5-hydroxymethyl cytosine plays the role that cytosine plays in other creatures. The DNA of a number of organisms contains both cytosine and methyl cytosine. In

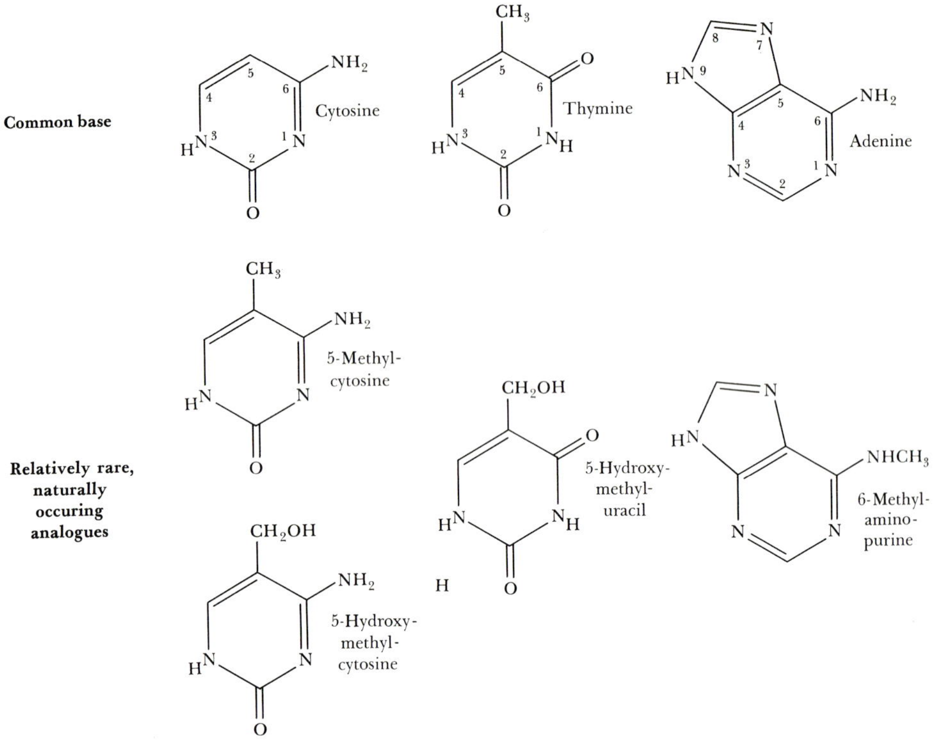

38 THE MECHANICS OF INHERITANCE

FIG. 2.15. *The structures of relatively rare nitrogenous bases that are known to occur in DNA from some sources. The structural formulas of the rare bases are compared with the formulas of the common bases for which they substitute. Note that the substitutions on the unusual bases avoid those positions involved in the adenine–thymine and guanine–cytosine hydrogen bonding (see Fig. 2.12).*

these cases, the sum of the amounts of these two bases is equal to the amount of guanine. A virus has been discovered that contains 5-hydroxymethyl uracil and another which contains uracil in place of thymine. The purine 6-methylamino purine occurs in some DNA's in low frequency in place of adenine. The structural formulas of these rare bases are shown and compared with their common analogues in Fig. 2.15. Other rare bases doubtless occur in nature. The special roles of some of them are partly known but cannot be discussed here.

The Structure of Viral RNA

*R*ibonucleic *a*cid (RNA) is cataloged by biologists according to the function it performs. In Chap. 10, three kinds of RNA that play key roles in the biosynthesis of proteins will be examined. We shall focus our attention here on RNA that plays a genetic role, that is, on "viral RNA."

FIG. 2.16. *A stretch of an RNA chain. Each of its four bases is attached to a "backbone" of ribose units connected by 3'-5' phosphate "bridges." The base sequence has been chosen willy-nilly.*

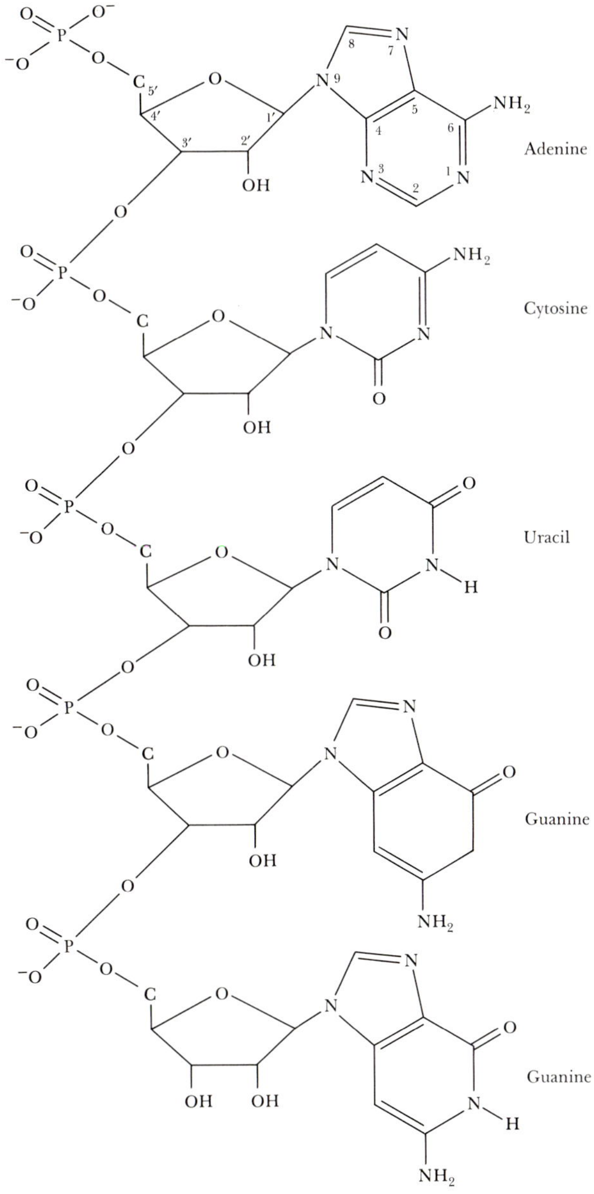

Adenine

Cytosine

Uracil

Guanine

Guanine

In all RNA the constituents are ribose (a 5-carbon sugar), phosphoric acid, and (usually) four nitrogenous bases. Although in some of the RNA's involved in protein synthesis a number of kinds of bases occur in small amounts in addition to the four predominant ones, in viral RNA only the bases adenine, guanine, cytosine, and uracil occur. Uracil is a pyrimidine identical to thymine except that the 5-position is not methylated.

The constituents of RNA are bonded together much as are the constituents of DNA. RNA is a long polymer with a sugar-phosphate backbone. The phosphate bridge connects the number 5 carbon atom of one ribose to the number 3 carbon atom of the next. The bases are attached to the ribose in exactly the same way the bases in DNA are attached to the deoxyribose. A stretch of an RNA chain is diagrammed in Fig. 2.16.

The RNA in the mature particles of most RNA-containing viruses is single-chained. This RNA becomes double-chained during the duplicative stage of the virus life cycle (see Chap. 3).

The chain length of viral RNA is of the order of 10^3 to 10^4 nucleotides, depending on the virus from which it is isolated.

Summary

The genic material of a number of bacteria and viruses is demonstrably DNA. In some of the smallest viruses RNA is the genic material. Circumstantial evidence from a variety of organisms leads to the hypothesis that, with the exception of those viruses, DNA is the universal genic material.

Chapters 3 and 4 will describe behavior of DNA in terms of its structure. Subsequent chapters will discuss genetic phenomena at some remove from the molecular level; in these chapters we shall share the aims of today's geneticist—to seek explanations for these phenomena in terms of the structure and properties of DNA.

References

Avery, Oswald T., Colin M. MacLeod, and Maclyn McCarty, "Studies on the Chemical Nature of the Substance Inducing Transformation of Pneumococcal Types," *J. Exp. Med., 79* (1944), 137–58. Reprinted in *Classic Papers in Genetics,* J. A. Peters, ed. (Englewood Cliffs, N.J.: Prentice-Hall, Inc., 1959), pp. 173–92, and in *Papers on Bacterial Genetics,* Edward A. Adelberg, ed. (Boston: Little, Brown & Co., 1960), pp. 147–68. The beauty of the experiments is equaled by the clarity of the exposition.

Fraenkel-Conrat, Heins L., and Robley C. Williams, "Reconstitution of Tobacco Mosaic Virus from Its Inactive Protein and Nucleic Acid

Components," *Proc. Natl. Acad. Sci. U.S., 41* (1955), 690–98. Reprinted in *Classic Papers in Genetics,* pp. 264–71. Tobacco mosaic virus that has been dissociated into its noninfectious protein and its slightly infectious RNA components can be reconstituted in vitro.

Hershey, A. D., and Martha C. Chase, "Independent Functions of Viral Protein and Nucleic Acid in Growth of Bacteriophage," *J. Gen. Physiol., 36* (1952), 39–56. Reprinted in *Papers on Bacterial Viruses,* G. S. Stent, ed. (Boston: Little, Brown & Co., 1960), pp. 87–104. "The experiments reported in this paper show that one of the first steps in the growth of T2 is the release from its protein coat of the nucleic acid of the virus particle, after which the bulk of the sulfur-containing protein has no further function." (From the authors' introduction.)

Stent, G. S., *Molecular Biology of Bacterial Viruses.* San Francisco: W. H. Freeman & Co., 1963. A "phage book" which has been cited as "an example of that rare item—a nearly perfect book."

Watson, James D., *The Double Helix.* New York: Atheneum Press, 1968. Perhaps the only reference you'll really enjoy.

———, and Francis H. C. Crick, "Molecular Structure of Nucleic Acids: A Structure for Deoxyribose Nucleic Acid," *Nature, 171* (1953), 737–38. Reprinted in *Classic Papers in Genetics,* J. A. Peters, ed. (Englewood Cliffs, N.J.: Prentice-Hall, Inc., 1959), pp. 241–43. The announcement of the Watson–Crick model for the structure of DNA.

Wilkins, Maurice, "Physical Studies of the Molecular Structure of Deoxyribose Nucleic Acid and Nucleoprotein," *Cold Spring Harbor Symp. Quant. Biol., 21* (1956), 75–90. Continued study following the Watson–Crick announcement served to substantiate their model while refining some of the dimensional parameters.

Problems

2.1. Calculate the molecular weights of (*a*) cytosine, (*b*) thymine, (*c*) guanine, and (*d*) adenine. Calculate the molecular weights of the corresponding nucleosides: (*e*) deoxycytidine, (*f*) (deoxy)thymidine, (*g*) deoxyguanosine, and (*h*) deoxyadenosine. Calculate the approximate molecular weight of (*i*) the sodium salt of a DNA molecule containing 10^4 base pairs in which the two possible base pairs are assumed to be about equally frequent.

2.2. The phage T2 contains not more than 2×10^5 nucleotide pairs. How long is the total DNA complement of T2? Compare your answer with the dimensions of a mature T2 particle (Fig. 2.5).

2.3. How many twists (approximately) are in a DNA molecule whose molecular weight is 10^8?

2.4. Calculate the number of possible sequences that can conceivably exist in a single-chain DNA molecule 10^4 nucleotides long. Assume no restrictions on the relative frequencies of the four kinds of bases.

2.5. Diagram the bit of DNA in Fig. 2.13 using A, T, C, and G for the bases, S for the sugar, and (P) for the phosphates. Indicate the end of each chain which bears a 3' —OH group.

Three

Duplication of Nucleic Acid

It was once a favorite hunch in biology that genetic specificity might be transmitted to proteins as the contours of a mold are transferred to the statue made in it; for instance, it was imagined that the genic material could direct the synthesis of compounds structurally complementary to itself. Such a notion is appealing both to naïve intuition and to the principles of structural chemistry. However, the idea is a bit demanding biologically; it insists that genic material can conduct the synthesis of both complementary molecules (genic products) and identical molecules (new genic material). Perhaps it was partly because of such considerations that the self-complementarity of the DNA molecule suggested to Watson and Crick a model for the duplication of DNA.

Crick and Watson hypothesized that prior to (or during) duplication the two chains of a DNA molecule separate. Each chain then directs the synthesis of a chain complementary to itself. It is usually imagined that synthesis of a new chain starts at one end of an old chain and proceeds step by step to the other end. The manner in which the polymerization is initiated is not easily visualized, but, once started, the rules for its continuation seem simple. The growing end of a new chain and the not-yet-copied portion of the old chain form a "corner." The corner formed at any particular level can be comfortably occupied by only one of the four nucleotides.

Only that nucleotide will be incorporated that can simultaneously form a hydrogen bond with the old chain *and* present its phosphate group at the proper position to be bonded to the growing end of the new chain. In a later section of this chapter we shall look at the mechanism of the bonding of each nucleotide to the growing end of a new chain. For now, let us consider the Watson–Crick scheme in terms of its success in predicting the rules of transmission of atoms from a parental DNA duplex into the two daughter molecules.

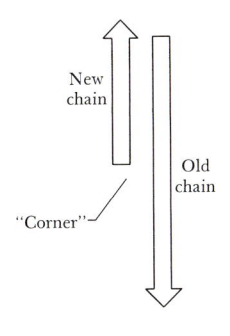

"Semiconservative Transmission" of Isotopic Label

A feature of the Watson–Crick hypothesis for DNA duplication is that half of each molecule ends up in a daughter molecule as a consequence of duplication; that is, duplication is "semiconservative." The prediction of semiconservative duplication implicit in the Watson–Crick hypothesis was confirmed by the outcome of an experiment performed by M. S. Meselson and myself in 1958 (see Fig. 4.1). We grew bacteria (*E. coli*) for many generations in a simple medium in which the only nitrogen source was NH_4Cl. The ammonium chloride contained only the heavy isotope of nitrogen, N^{15}. A preponderance of the ordinary isotope of nitrogen, N^{14}, was then added to the medium. At intervals cells were removed, and the DNA was extracted from them. The density of the individual molecules in each sample was then examined by the method of equilibrium density gradient centrifugation to determine the relative content of N^{15} and N^{14}.

In the equilibrium density gradient method a concentrated salt solution (for example, cesium chloride, CsCl, in water) containing the DNA to be examined is centrifuged in an analytical ultracentrifuge. The CsCl, since it is more dense than water, tends to sediment to the outside of the cell. This tendency to sediment is opposed by diffusion, however, and after about 8 hr (under the conditions employed) the concentration distribution of CsCl in the cell is essentially stable. The resulting smooth concentration gradient results in a density gradient. The DNA molecules in the cell are driven to the region in the gradient that corresponds to their own effective density. The bands into which the DNA collects are of finite width, since DNA, too, is subject to diffusion. By the end of about 20 hr the sedimentation and diffusion forces acting on the DNA are essentially in equilibrium, and the concentration of DNA can be photographically determined. The principal result of the experiment was that after one doubling of the *E. coli* population each molecule contained equal amounts of N^{15} and N^{14}.

44   THE MECHANICS OF INHERITANCE

FIG. 3.1. (a) *Ultraviolet absorption photographs showing DNA bands resulting from density gradient centrifugation of lysates of bacteria sampled at various times after the addition of an excess of N^{14} substrates to a growing N^{15}-labeled culture. The density of the CsCl solution increases to the right. Regions of equal density occupy the same horizontal position on each photograph.* (b) *Microdensitometer tracings of the DNA bands shown in the adjacent photographs. The microdensi-*

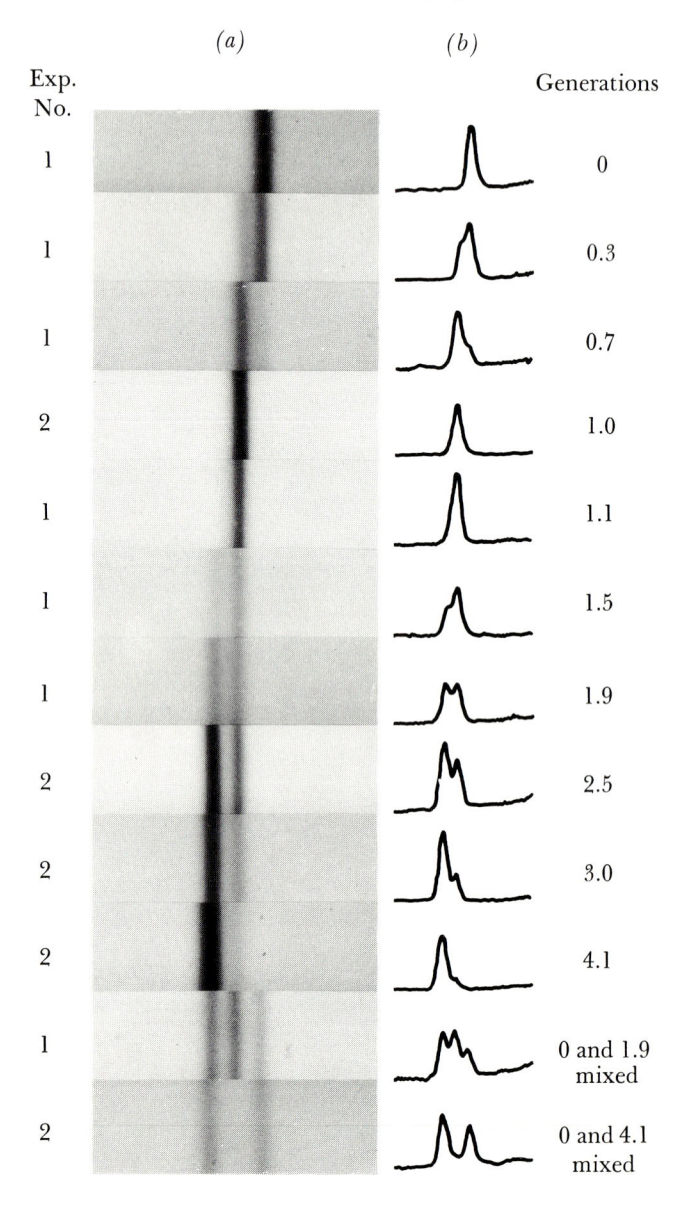

Exp. No.	(a)	(b)	Generations
1			0
1			0.3
1			0.7
2			1.0
1			1.1
1			1.5
1			1.9
2			2.5
2			3.0
2			4.1
1			0 and 1.9 mixed
2			0 and 4.1 mixed

Duplication of Nucleic Acid 45

tometer pen displacement above the base line is directly proportional to the concentration of DNA. The degree of labeling of a species of DNA corresponds to the relative position of its bands between the bands of fully labeled and unlabeled DNA in the lowermost frame, which serves as a density reference. This figure, reproduced with permission of M. S. Meselson, appeared in Proc. Natl. Acad. Sci. U.S., **44** *(1958), 675.*

FIG. 3.2. *The distribution of parental atoms to daughter DNA molecules. At the first duplication the two isotopically labeled strands of the "original parent molecule" are transmitted one into each daughter molecule. Upon continued duplication, the two original parent chains remain intact, so that there will always be found two molecules, each with one labeled chain. The arrowheads indicate the chemical polarity of the DNA chains. This figure, reproduced with permission of M. S. Meselson, appeared in* Proc. Natl. Acad. Sci. U.S., **44** *(1958), 678.*

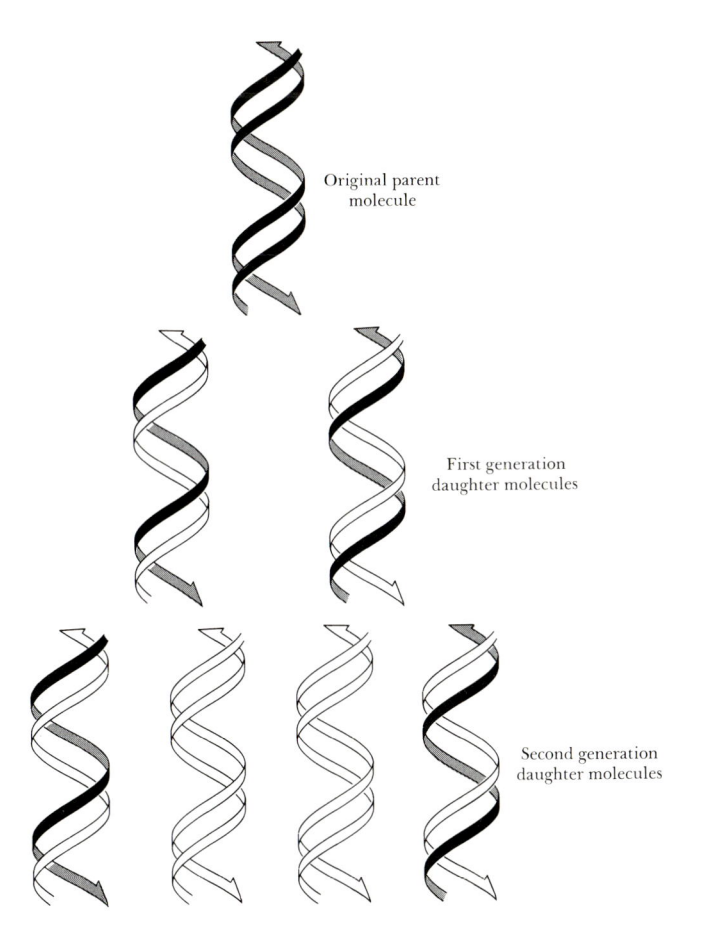

Original parent molecule

First generation daughter molecules

Second generation daughter molecules

After two doublings, there were present equal numbers of such "half-labeled" molecules and unlabeled molecules, containing only N^{14} (see Fig. 3.1). The results of this experiment are schematically summarized in Fig. 3.2.

It is clear from the foregoing that the net result of DNA duplication is that two separated old chains each are found associated with their complementary new chains. This result, which has been found to hold for DNA from many sources (including humans and plants), is precisely the prediction made by Watson and Crick. Herbert Jehle has put forth a modification of the Watson–Crick scheme, however, which leads to the same rules for the distribution of parental atoms but which deviates from the notion that each old chain directs the synthesis of the new chain with which it is associated in the daughter molecule. In Jehle's scheme, diagrammed in Fig. 3.3, one of the two parental

FIG. 3.3. *Duplication of DNA according to the scheme of Herbert Jehle. In front (a) of the region of duplication, the two old chains (white) are associated in a Watson–Crick duplex (indicated by the light shading between them). At point b a new chain (shaded) is growing downward, using one of the old chains as a template and displacing the other old chain as it goes. At point c a second new chain is growing downward, using the first new chain as a template. At d the chains are sorted out semiconservatively.*

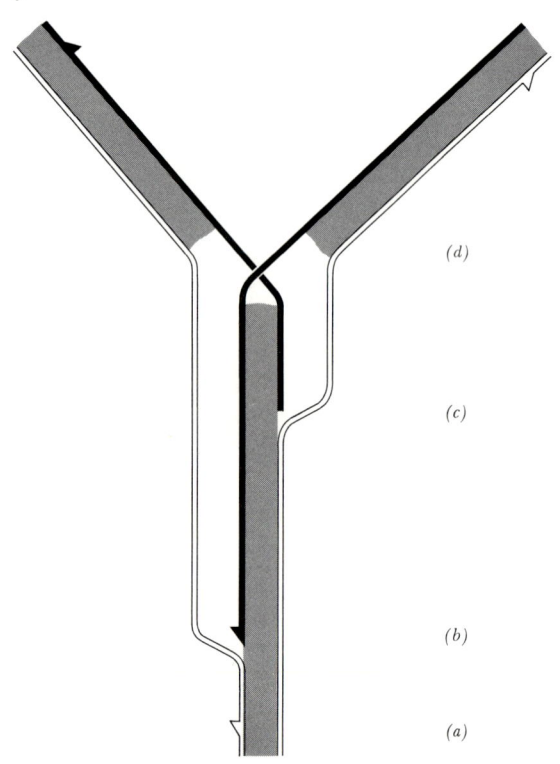

Duplication of Nucleic Acid 47

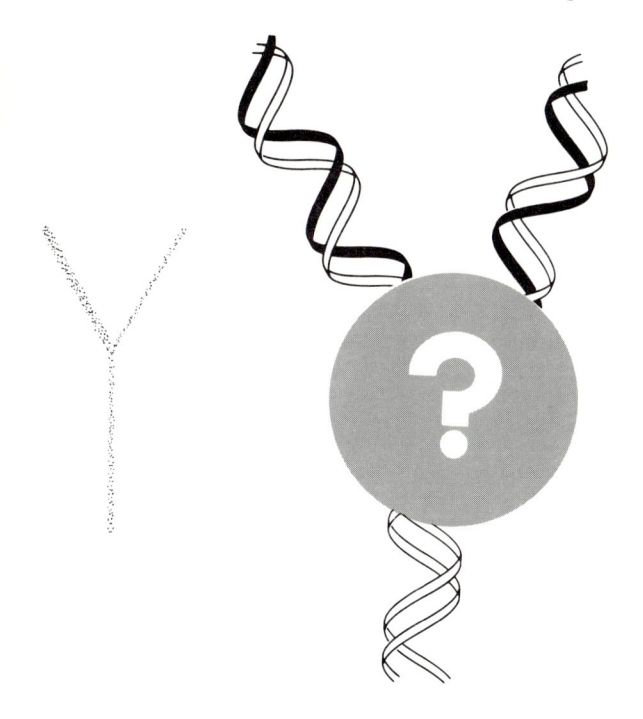

FIG. 3.4. *During DNA duplication, the synthesis of new chains (shaded) goes on hand in hand with the unwinding of the two old chains (not shaded). See text for an explanation of the "bird track" on the left.*

chains directs the synthesis of a chain complementary to itself; the new chain then acts as a template for the production of a chain complementary to *itself*. The result is two new chains of opposite polarity which are then assorted one into each daughter molecule. In the following chapter we shall consider a particular consequence of the notion of synthesis directed by complementary base pairing; we will then be in a position to consider experiments which might distinguish the Watson–Crick model from Jehle's modification of it. At the moment, another feature of DNA duplication, consonant with both models, seems clear—chain synthesis does go on hand-in-hand with chain separation. A molecule in the act of duplication "looks like" the one in Fig. 3.4. John Cairns established this view by examining autoradiographs (see legend to Fig. 5.3) of fragments of DNA from *E. coli* that had been allowed to duplicate for varying periods of time in radioactive thymidine. In molecules that were undergoing their second round of duplication in radioactive medium, a region could be seen that gave an autoradiographic image like that shown in the small drawing to the left of Fig. 3.4. Under the experimental conditions employed, the density of ion tracks in the image is proportional to the

"Plus" chain + {"ATP", "UTP", "GTP", "CTP"}

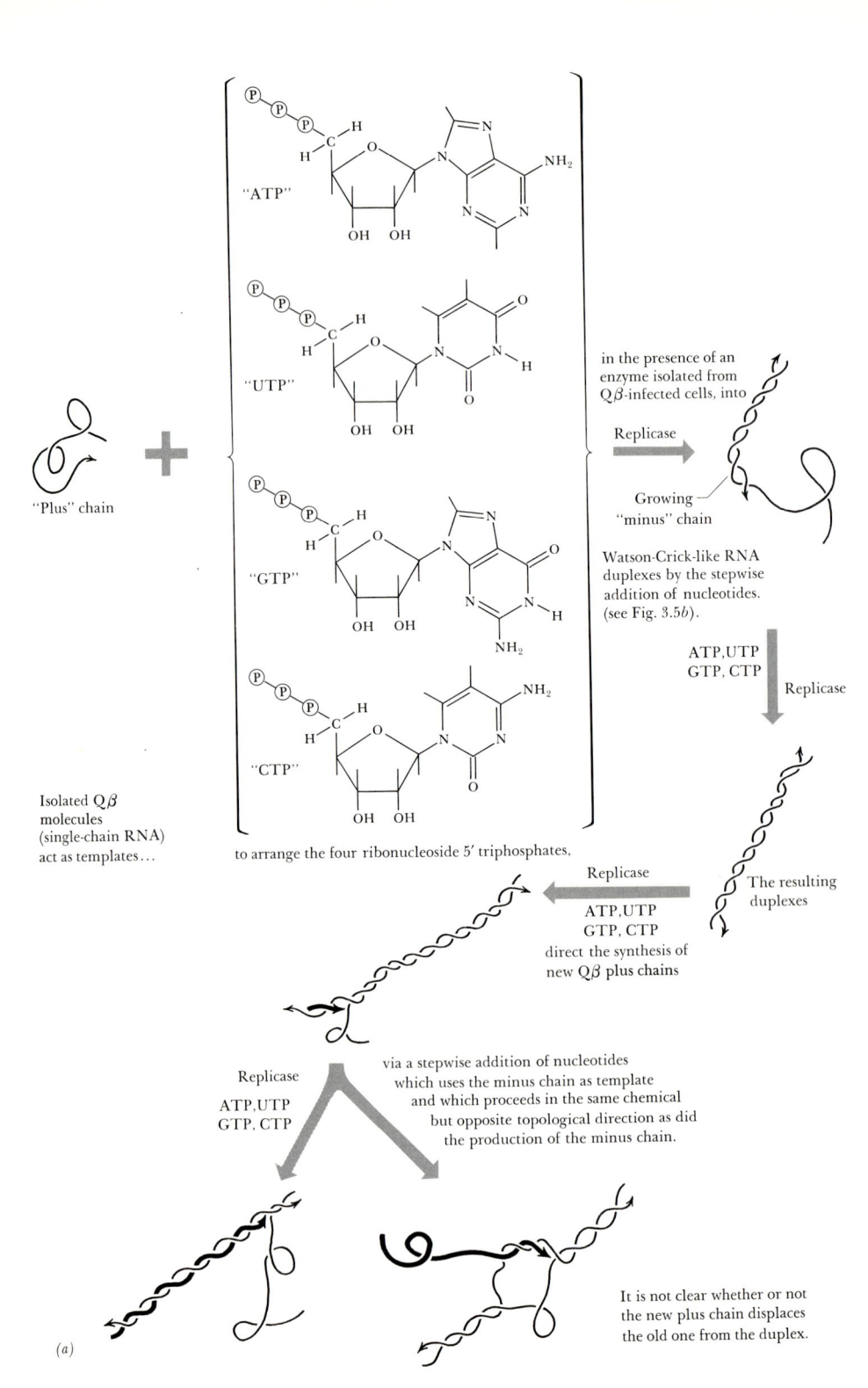

Isolated Qβ molecules (single-chain RNA) act as templates... to arrange the four ribonucleoside 5' triphosphates, in the presence of an enzyme isolated from Qβ-infected cells, into

Replicase →

Growing "minus" chain

Watson-Crick-like RNA duplexes by the stepwise addition of nucleotides. (see Fig. 3.5b).

ATP,UTP GTP, CTP
Replicase

The resulting duplexes

← Replicase
ATP,UTP GTP, CTP

direct the synthesis of new Qβ plus chains

Replicase
ATP,UTP GTP, CTP

via a stepwise addition of nucleotides which uses the minus chain as template and which proceeds in the same chemical but opposite topological direction as did the production of the minus chain.

It is not clear whether or not the new plus chain displaces the old one from the duplex.

(a)

number of labeled chains at any level in the molecule. Convince yourself that the pattern just described would result if the half-labeled molecules arising as a result of one generation in labeled medium were undergoing their second duplication in labeled medium in the fashion diagrammed in Fig. 3.4.

In Vitro Synthesis of Genic Nucleic Acid

The syntheses of both DNA and genic RNA have been accomplished in vitro. The study of these "life in a test tube" reactions is certain to advance profoundly our understanding of the mechanism(s) of duplication of genic material. The in vitro systems require the fol-

FIG. 3.5. (a) Facing page. *The duplication of the RNA molecule of virus Qβ as revealed by in vitro studies.* (b) *The addition of nucleotides to a growing nucleic acid chain. The geometry of nucleotide addition is shown and described early in this chapter. The chemistry of the addition involves the splitting of pyrophosphate (P–P) from the incoming nucleoside 5′-triphosphate and the formation of an ester bond between the remaining phosphate group and the 3′ —OH group of the terminal nucleotide. The reaction is diagrammed above for the case of the addition of uridylic acid to the end of a growing RNA chain.*

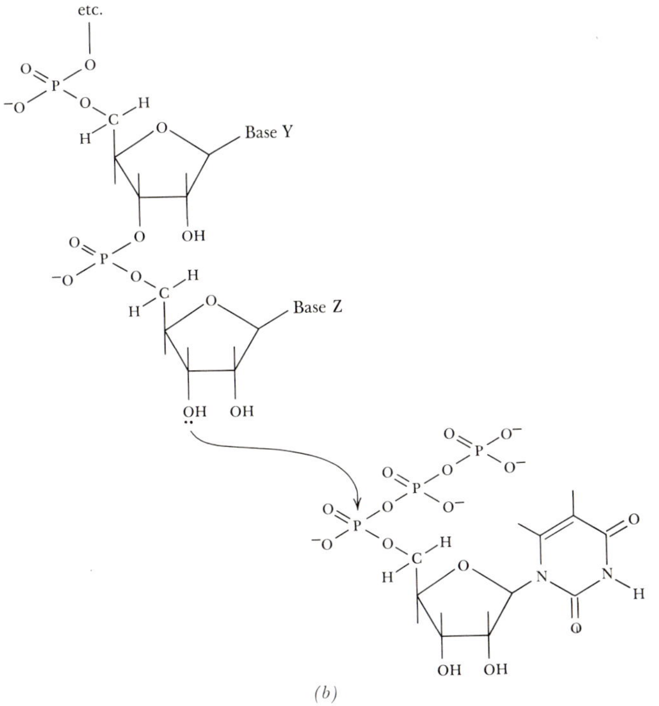

(b)

lowing three components for synthesis to proceed: (1) a nucleic acid template; (2) nucleoside 5'-triphosphates (see Fig. 3.5); and (3) an enzyme which connects the monomers into sequences dictated by the template. Taking first things first, let's look at S. Spiegleman's in vitro duplication of viral RNA.

The phage $Q\beta$ is a small particle whose one RNA molecule is single-chained. Following its penetration into a host cell the RNA molecule directs the production of an enzyme ("replicase") which, when isolated, has the following properties: (1) In the presence of ribonucleoside 5'-triphosphates and intact $Q\beta$ RNA molecules (but no other kind of RNA!) the enzyme catalyzes the formation of RNA. This RNA includes predominantly molecules which are indistinguishable from the $Q\beta$ RNA molecules which were initially introduced. In particular, the newly produced RNA can act as primer for the production of more RNA, and the newly produced RNA can invade cells and direct the production of new, complete, infectious $Q\beta$ particles. Thus, there is ample reason to suppose that this "in vitro system" is a good mock-up of the in vivo replication of $Q\beta$ genic material. As such, it deserves the closest attention.

The sequence of steps in the in vitro production of $Q\beta$ RNA has been partially established. First, the single-chain RNA molecule (call it a "plus chain") acts as a template for the construction of a chain complementary to itself (a "minus chain"). The RNA molecule is thereby converted into a duplex. The minus chain then acts repetitively as a template for the construction of new plus chains (that is, new $Q\beta$ RNA molecules). These steps are summarized somewhat speculatively in Fig. 3.5.

By means of a rather involved sequence of manipulations, single-chain viral DNA has recently been produced in vitro. The intensive preliminary work with the components of this system was pioneering work in the in vitro synthesis of nucleic acid. In fact, the experiment on nearest-neighbor frequencies described in Chap. 2 was carried out in vitro with components of this system, developed by A. Kornberg. The degree to which this DNA-synthesizing system reflects in vivo events is uncertain. In vitro, the following steps are carried out: Single-chain DNA from a small phage plus density-labeled deoxynucleoside 5'-triphosphates are incubated with an enzyme, "DNA-polymerase," isolated from *E. coli*. The single-chain DNA becomes converted to duplexes. The chains of the duplexes are separated by heating, which "melts" the hydrogen bonds, and the dense minus strands are isolated by centrifugation in a CsCl density gradient. These minus chains can then act as templates for the production of new viral plus chains. Phage DNA molecules made (essentially) in this way are able to infect bacteria under appropriate conditions, causing the formation of populations of complete infectious phage particles.

Summary

The self-complementarity of duplex DNA suggested that duplication is a consequence of the production of new DNA chains upon templates of complementary structure (and opposite chemical polarity). The study of in vitro synthesis of nucleic acids confirms this notion. As elaborated by Watson and Crick, this notion predicted the semiconservative distribution of parental atoms of a DNA duplex into daughter molecules, a prediction which has been borne out for DNA from a wide variety of sources. In the following chapter, after we examine ideas and experiments regarding the origin of mutations, we can examine further consequences of the Watson–Crick scheme for DNA duplication.

References

Cairns, John, "The Bacterial Chromosome and Its Manner of Replication as Seen by Autoradiography," *J. Mol. Biol., 6* (1963), 208–13. The replication of DNA really *does* go hand in hand with chain separation.

Goulian, M., A. Kornberg, and R. L. Sinsheimer, "Enzymatic Synthesis of DNA, XXIV. Synthesis of Infectious Phage ϕx174 DNA," *Proc. Natl. Acad. Sci., 58* (1967), 2321–28. DNA-life-in-a-test-tube.

Jehle, Herbert, "Replication of Double-Strand Nucleic Acids," *Proc. Natl. Acad. Sci., 53* (1965), 1451–55. An interesting variation on the Watson–Crick duplication model.

Meselson, M. S., and F. W. Stahl, "The Replication of DNA in *Escherichia coli*," *Proc. Natl. Acad. Sci. U.S., 44* (1958), 671–82. Duplication of DNA is shown to be semiconservative, as predicted by Watson and Crick.

Spiegleman, S., I. Haruna, N. R. Pace, D. R. Mills, D. H. L. Bishop, J. R. Claybrook, and R. Peterson, "Studies in the Replication of Viral RNA," *J. Cell. Physiol., 70:* Suppl. (1967), 35–64. RNA-life-in-a-test-tube, a review with emphasis on the author's work.

Watson, J. D., and F. H. C. Crick, "The Structure of DNA," *Cold Spring Harbor Symp. Quant. Biol., 18* (1953), 123–31. Includes the Watson–Crick proposal for the mechanism of DNA duplication.

Problems

3.1. Consider a DNA molecule in which every nitrogen atom is "labeled," that is, in which every nitrogen atom is the heavy isotope (N^{15}) instead of the normal isotope (N^{14}). Imagine that such a molecule is permitted to duplicate in an environment in which all of the nitrogen is N^{14}.

(*a*) After one duplication, how many of the (two) molecules present contain some N^{15}?

52 THE MECHANICS OF INHERITANCE

(b) After three duplications, how many of the (eight) molecules present contain some N^{15}?

(c) Suppose that, after two duplications, the (four) molecules are returned to an environment containing only N^{15}. After one duplication in the N^{15} environment, how many of the (eight) molecules will contain some N^{14}?

(d) If each of the molecules in Problem 3.1a were cut transversely (as one cuts a piece of string) into a number of pieces, how many of the resulting pieces would contain some N^{15}? (This experiment has been done using ultrasonic vibrations to cut the DNA. I trust you deduced the same answer that was found experimentally.)

3.2. Nowadays, one laboratory uses N^{15} and C^{13} simultaneously in "DNA-transfer" experiments. What do you calculate is the approximate density difference between "heavy" and "light" DNA in this case? (Assume equal frequencies of adenine and guanine.)

3.3. Duplication of phage DNA begins about 6 min after a T2-phage particle attacks an *E. coli* cell. Eleven min or so after the moment of infection about 30 (say 32) phage-equivalents of DNA are present in the cell, and mature particles begin to appear, their number increasing linearly with time. DNA synthesis continues at a constant rate equal to the rate of maturation, so that the amount of "naked" DNA remains constant. At about 22 min, when the average number of mature phage is approximately 200, the cell stops producing phage and falls apart (lyses). Assume that phage chromosomes multiply as do bacteria, for instance, one gives two, each of which can double to give four, and so on.

(a) Calculate the average "generation time" for T2 DNA during the period from 6 min to 11 min after infection. (In Chap. 5 we shall examine evidence indicating that the DNA of a T2 particle is one molecule about 2×10^5 nucleotide pairs long.)

(b) What is the rate of synthesis of T2 DNA in nucleotide pairs per minute?

3.4. You have an RNA phage whose chromosome contain bases in the following proportions: adenine 25 percent, guanine 15 percent, uracil 20 percent, cytosine 40 percent. In an in vitro system these chromosomes are converted to duplexes. What are the relative amounts of the four bases in these duplexes?

3.5. A biochemist synthesized DNA in vitro using Kornberg's DNA polymerase and a duplex template in which the AT/GC ratio is not equal to 1. In Experiment No. 1 the nucleoside triphosphate of adenine used as precursor material was labeled with P^{32}. In Experiment No. 2, the nucleoside triphosphate of thymine was so labeled. The nucleoside triphosphates of guanine and cytosine were so labeled in Experiments Nos. 3 and 4, respectively. The radioactive DNA isolated from each experiment was enzymatically hydrolyzed by a DNAase which cleaves the ester bonds connecting 5'-carbon atoms to phosphorus.

The biochemist measured the amounts of P^{32} in some of the 3'-nucleotides isolated from each DNA preparation. Those amounts are entered in the table below. He figured that he did not have to measure them *all*, because he believes in Watson and Crick and the Kornberg enzyme.

(*a*) Fill in the blanks in the accompanying table.

Experiment	1	2	3	4
Precursor nucleoside 5'-triphosphate	A	T	G	C
Recovered 3'-nucleotide				
A	20	20	40	20
T	40			
C	30		55	
G	10		25	

The table gives the relative amounts of P^{32} recovered in the indicated 3'-nucleotides.

(*b*) What is the AT/GC ratio for this DNA?

Four

Mutation of DNA

The hypothesis for the mechanism of DNA duplication discussed in Chapter 3 was suggested to Watson and Crick by the molecular structure of DNA. The duplication scheme in turn suggested to its authors a mechanism of mutation.

Tautomerism of Bases

Each of the four bases in DNA can exist in states alternative to those shown in the previous figures. The transitions to these rare states occur by rearrangements (tautomeric shifts) in the distribution of electrons and protons in the molecule. The (interesting) tautomeric alternatives for each of the four bases are shown in Fig. 4.1. When a base is in its rare state, it cannot form a hydrogen-bonded pair within a DNA molecule with its usual partner. A purine in its rare state can, however, form a fine pair with the "wrong" pyrimidine, and a pyrimidine in its rare state can form a fine pair with the "wrong" purine. The four pairings that can occur when one member (but not both) of a base pair has undergone a tautomeric shift are diagrammed in Fig. 4.2.

Watson and Crick suggested that the occurrence of the rare tautomeric alternatives for each of the bases provides a mechanism for mutation during DNA duplication. If

Mutation of DNA 55

FIG. 4.1. *Tautomerism of the bases of DNA. Structural formulas for each of the four bases of DNA can be written in several alternative ways. The alternatives differ from each other only by rearrangements of protons and electrons. These rearrangements are symbolized in the structural formulas by changes in the positions of H atoms and of double bonds. Any rearrangement involving only protons and electrons may be written as long as the valences of the constituent atoms are respected. The frequencies of occurrence of the rare tautomers are so low (<<1%) that they cannot be accurately determined. The bases of DNA occur overwhelmingly in the states shown at the left. The tautomer suspected to be important in the origin of mutations is shown at the right.*

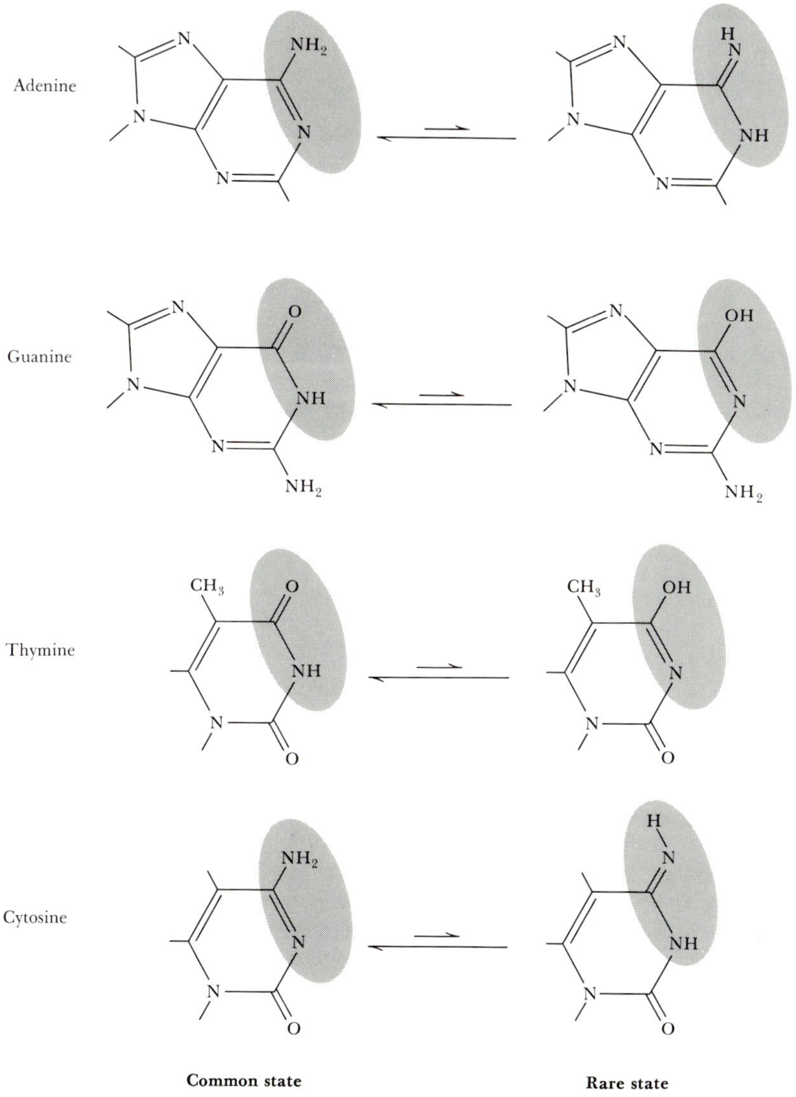

Adenine

Guanine

Thymine

Cytosine

Common state **Rare state**

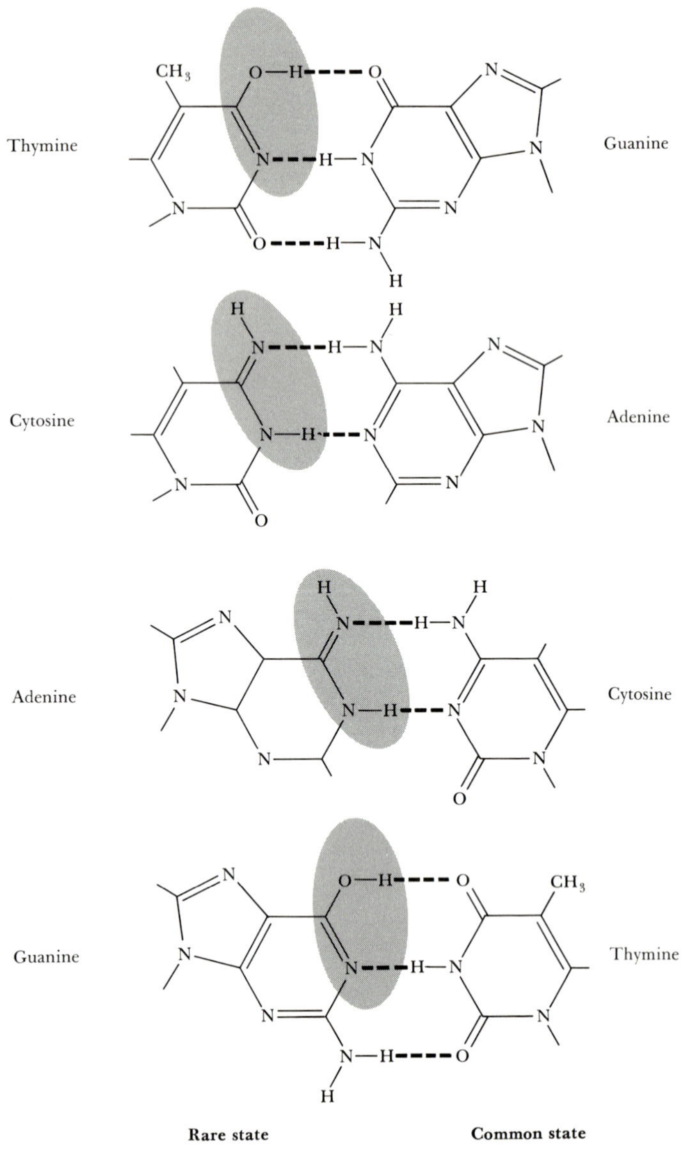

56 THE MECHANICS OF INHERITANCE

FIG. 4.2. *"Forbidden base pairs" resulting from tautomerization. The specific hydrogen-bonding properties of each of the bases are reversed by the tautomeric shifts in Fig. 4.1.*

Mutation of DNA   57

FIG. 4.3. *The consequences of tautomerization of an adenine residue at the time of DNA duplication. Tautomerization can be instrumental in the induction of base-pair transitions in two ways: (1) A base, adenine in the case depicted here, already in a DNA chain may tautomerize at the moment a new DNA chain is being synthesized along it. (2) Adenine deoxyribose triphosphate about to be incorporated at the growing end of a new chain may tautomerize. Note that these two cases have opposite results. The former case induces the transition AT to GC while the latter induces the transition GC to AT. Note also that for each case the first product of the mistake is a heteroduplex. In the Watson–Crick model for duplication the heteroduplexes give rise in the next duplication to a molecule of the original type as well as to a mutated molecule.*

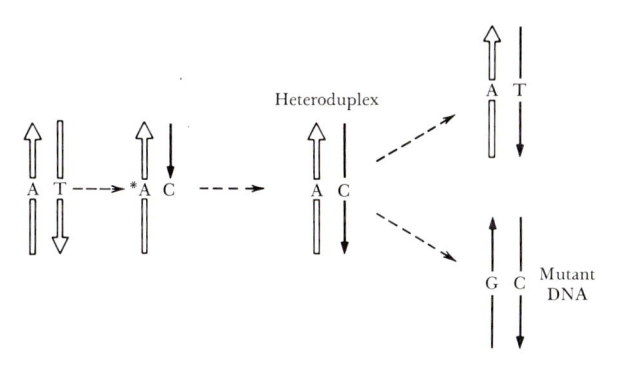

Mutation by tautomerism of incorporated adenine (above)

Mutation by tautomerism of incoming adenine (below)

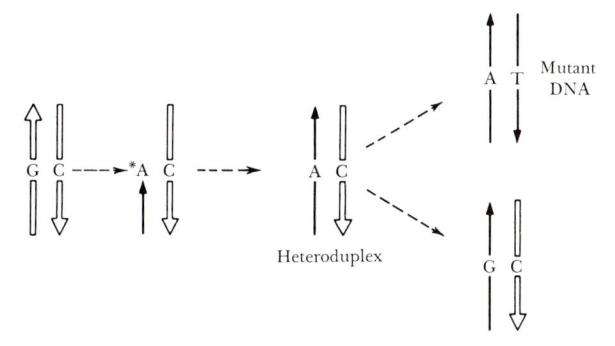

a base in an old chain is in its rare form at the moment that the growing end of the complementary new chain reaches it, a wrong nucleotide can be added to the growing end. Similarly, if the base of a nucleoside triphosphate is in *its* rare form, it may be added to the growing end of a new chain at an incorrect level. In either case, the primary consequence of this mistake is the formation of a DNA molecule that contains a "forbidden" base pair. In the Watson–Crick model for duplication, such a heteroduplex would give rise upon duplication to two kinds of DNA molecules—one is identical to the original DNA; the other has undergone a base-pair substitution at the level of the pairing mistake induced by the tautomeric shift. Within Jehle's scheme for DNA duplication what would be the products of duplication of a heteroduplex? Later on in this chapter we shall look at some experimental situations that relate to this difference between the two duplication models.

Substitutions of the sort predicted by Watson and Crick, for example, $AT \rightleftharpoons GC$ or $TA \rightleftharpoons CG$, are called transitions. They have the feature that a purine on one chain is replaced by a different purine, and a pyrimidine on the other chain is replaced by a different pyrimidine. Other base substitutions are logically possible. These all involve the replacement of a purine by a pyrimidine on one chain and a pyrimidine by a purine on the other. These changes are called transversions. Transversions do occur, and in Chap. 10 we shall discuss the kind of evidence that indicates their occurrence. All the possible base substitutions are summarized in the accompanying scheme.

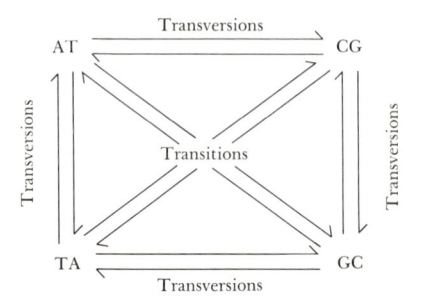

The steps in transition by tautomerism are outlined in Fig. 4.3.

It has been suggested that transitions can also arise as a consequence of ionization of a base at the time of DNA duplication. The loss of the proton from the number 1 nitrogen of either thymine or guanine permits the formation of a TG base pair. The base-pairing configurations that become permissible because of these ionizations are diagrammed in Fig. 4.4.

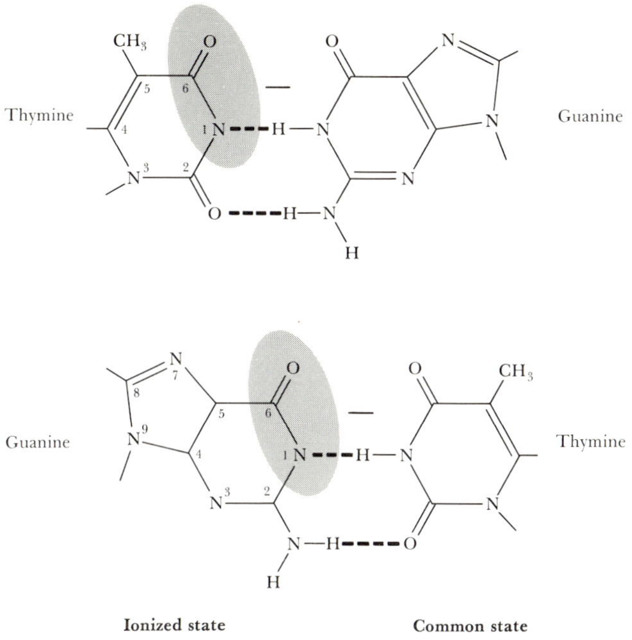

FIG. 4.4. *"Forbidden base pairs" resulting from ionization of the number 1 nitrogen. The loss of the proton from the number 1 nitrogen of thymine or guanine permits the formation of the base pair TG. The double bond from the number 6 carbons to the oxygens could as well be drawn between the 6 and 1 positions of the rings.*

Mutation by Base-Pair Transitions

The notion of mutagenesis as a result of base-pairing mistakes at the time of DNA duplication has received considerable experimental support. Much of the support has come from studies of *rII* mutants of the *coli* phage T4. The discussion that follows is not meant to create the impression that mutation is fully understood, or even that what understanding we have rests primarily on studies of the *rII* system; but the clarity of concept and the simple consistency of results that have characterized many of the experiments and most of the publications dealing with *rII* make it useful for illustration.

T4*rII* particles differ from T4*r*+ (wild-type) particles in two useful respects. An appreciation of the usefulness of this experimental material requires short digressions into technique and history.

TECHNIQUE

Phage particles are conveniently enumerated by distributing a measured volume of a phage suspension upon the surface of a solid nutrient-agar medium along with about 100 million host bacteria.

60   THE MECHANICS OF INHERITANCE

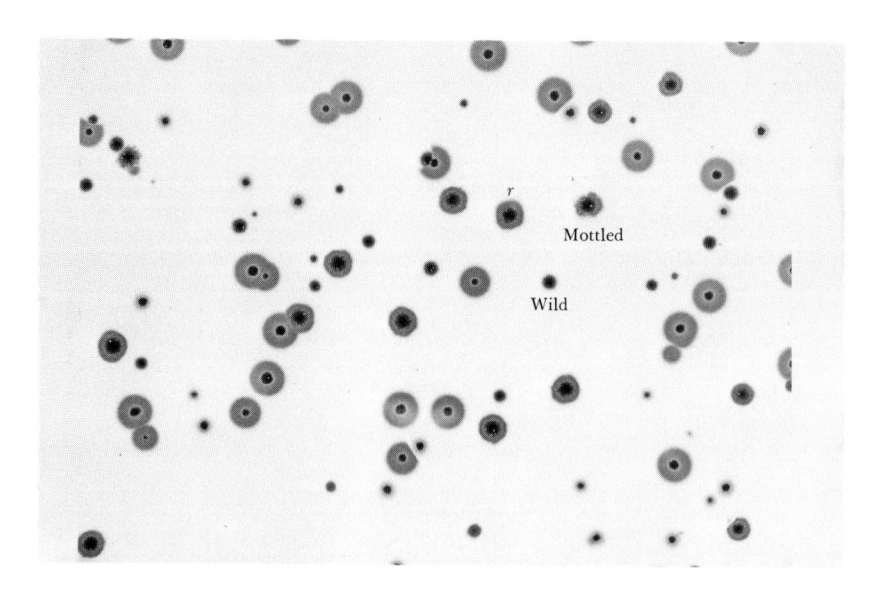

FIG. 4.5. *Photograph of T4-phage plaques. The light background is a "carpet" of* E. coli *cells; the dark spots (plaques) are regions in which cells have been destroyed by the multiplying viruses. Each plaque arose from a single particle. Eight different hereditary types of T4 generated these plaques. The plaques of most interest to us in this chapter are wild type and* r. *The mottled plaque is the sort of plaque that results from growth of a heteroduplex (see page 65). This photograph was made by using the petri plate itself in place of a negative in an ordinary photographic enlarger.*

This is accomplished by suspending the phages and cells in a few ml of melted nutrient agar and then pouring the inoculated agar onto the surface of the solid agar medium. The melted agar spreads across the surface and then, within a few minutes, hardens to a semisolid consistency. The bacteria multiply to form an array of colonies so dense as to constitute a uniform sheet of bacterial cells in the top agar layer. Each of the phage particles adsorbs to, multiplies within, and lyses a nearby bacterium. Upon their release, the progeny particles adsorb to neighboring cells and repeat the cycle. Within a few hours a number of holes (plaques) corresponding to the number of particles introduced into the top layer agar have been eaten in the otherwise continuous sheet of bacterial cells.

T4*rII* particles, when grown with *E. coli* strain B, make plaques that are larger and have a sharper edge than do those of T4*r*+ (see Fig. 4.5). When grown with *E. coli* K12(λ),[1] T4*r*+ particles make plaques as they do on strain B; T4*rII* particles, however, fail to make plaques on K12(λ).

[1] *Escherichia coli* strain K12 carrying the prophage λ (see Chapter 8).

HISTORY I

S. E. Luria showed that mutation rates in bacteriophages can be determined by application of the same equations used to measure mutation rates in bacterial cultures (see Chap. 1). An implication of Eqs. 1.1 and 1.3 is that the number of mutations occurring in the last generation of a bacterial culture is equal to the sum of the number occurring in all the generations previous to that one. These same equations imply that the generation during which a mutation occurs can be determined by the number of (mutant) offspring deriving from the mutational event. A mutation in the ith generation gives rise to twice as many mutants in a culture as one occurring in the ith $+$ 1 generation; a mutation in the last generation leaves but one (mutant) offspring.

Luria examined the phage progeny liberated by individual bacteria infected by r^+ phages. The average phage yield per cell was about 80 particles. Of those cells that yielded some r particles, half yielded only one r particle; the other half yielded two or more. Similarly, one-fourth of the bursting cells yielded four or more mutant particles and one-eighth yielded eight or more, and so on. This observation clearly implies that phages multiply "exponentially" (as do bacteria) and justifies the use of the same definition for mutation rate in phages as is used for bacteria.[2]

HISTORY II

S. Benzer recognized the virtues of the T4rII mutants and set an extraordinary example of vigorous exploitation.

Thousands of rII mutations have been isolated from independent occurrences of mutation. Most of these mutants are distinguishable from each other by one or both of two criteria. (1) Their rates of back mutation to r^+ may be distinguishable. (2) When grown *together* in a host cell they may produce r^+ progeny in a higher frequency than when either strain is grown by itself. (This observation, without further explanation, serves to establish the nonidentity of the two mutant strains. The nature of the interaction (genetic recombination) that leads to the production of r^+ progeny is described in Chap. 6 and 7.) Those of the rII mutants that arise *à la* Watson and Crick must be supposed to represent transitions of single base pairs. Furthermore,

---

[2] This discussion is a simplification. The two-step nature of the production of a mutant DNA molecule diagrammed in Fig. 4.3 is not incorporated in the algebra; neither is the multinucleate nature of many bacteria—a situation that can lead to analogous analytical ambiguities. For phages the situation is even more complicated (see Problem 3.3). Nevertheless, the growth equation (Eq. 1.1) and the definition of mutation rate (Eq. 1.3) are useful approximations.

the ability to distinguish most of the mutants from each other indicates that alterations of each of a large number of base pairs can be studied. In Chap. 7 we shall examine the evidence that each of these alterations influencing the *rII* phenotype occurs within a restricted region of the T4 DNA. At this point it is the uniqueness of many of the *rII* mutations that is important. This uniqueness, coupled with the (re)discovery of chemical mutagenesis in viruses, made possible an experimental challenge of the Watson–Crick hypothesis for the mechanism of mutation.

Base-Pair Transitions Induced By 5-Bromouracil and Nitrous Acid

5-Bromouracil (5BU) is a structural analogue of thymine (5-methyl uracil). When phages are grown in the presence of 5BU, they may incorporate it into their newly synthesized chains of DNA. Indeed, if the infected cells are prevented from making thymine, all of the positions normally occupied by thymine in the newly synthesized chains are instead occupied by 5BU. Phages grown under conditions that permit 5BU incorporation have a higher than normal mutation

FIG. 4.6. *The molecular basis of mutation induction by 5-bromouracil. 5-Bromouracil is structurally so similar to thymine that it can substitute for thymine in the biosynthesis of DNA. A consequence of this substitution is a high mutation rate. The greater degree of ionization, and perhaps of tautomerization, of 5BU probably accounts for its mutagenic properties. The two proposed mechanisms have in common the loss of hydrogen from the number 1 ring position.*

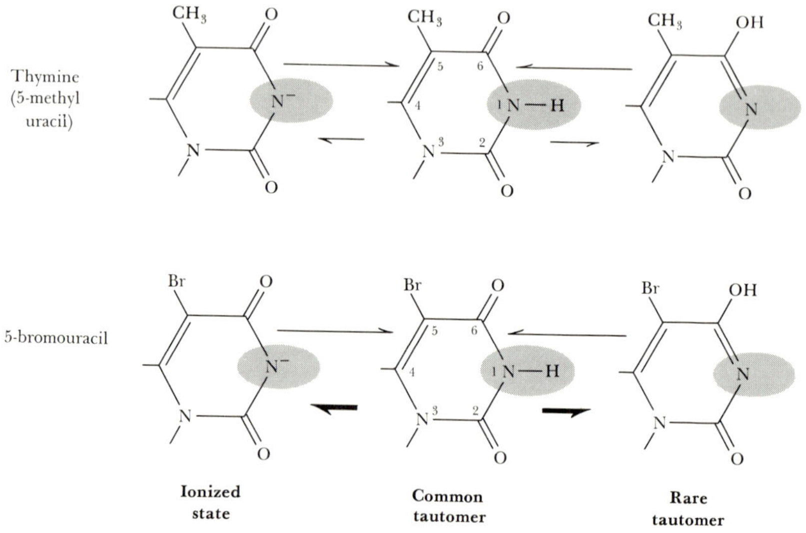

rate. "To be expected," say the philosophical heirs of Watson and Crick. From what we know of the effects of bromine substitution in cyclic organic compounds we should expect 5BU to have a rather high rate of migration of the proton from the number 1 ring nitrogen. Fig. 4.6 shows the tautomeric forms and the ionized forms of thymine and its analogue 5BU respectively. An inspection of Fig. 4.6 reveals that tautomerization and ionization have a structural consequence in common; they both involve the loss of hydrogen from the number 1 nitrogen. This common structural feature has evoked the prevalent proposals regarding their roles in mutagenesis.

The notion that 5BU effects mutations by inducing base-pair transitions is strengthened by several observations. We can discuss two of these now.

(1) Mutations induced in phages by duplication in the presence of 5BU can be made to back-mutate (revert) under the same conditions. This observation is easily understood if 5BU can induce both GC-to-AT and AT-to-GC transitions as described in Fig. 4.3 for spontaneous mutations arising from tautomerism of adenine.

(2) Among 5BU-induced mutants, some can revert only when 5BU is present as a precursor in DNA synthesis; others can revert when the only 5BU present is that already incorporated in DNA in place of thymine. Presumably the latter class represents transition of the type

$$\underset{mutant}{AT} \xrightarrow{\text{5BU in DNA}} \underset{wild\ type}{GC}$$

while the former class is of the other type:

$$\underset{mutant}{GC} \xrightarrow{\text{5BU precursor}} \underset{wild\ type}{AT}$$

Nitrous acid (HNO_2) treatment of $-NH_2$ substituted purines and pyrimidines results in replacement of the $-NH_2$ (amino) group by an $-OH$ (hydroxyl) group. The consequences of HNO_2 treatment of adenine and cytosine are diagrammed in Fig. 4.7. The deamination of cytosine leads to the formation of a product (uracil) whose hydrogen-bonding property are essentially those of thymine; the deamination of adenine results in the formation of hypoxanthine, whose hydrogen-bonding properties are similar to those of guanine. There are two reasons for thinking that the deamination of guanine is not mutagenic. (1) Xanthine, the product of deamination of guanine, has hydrogen-bonding properties like those of guanine, rather than like those of adenine. (2) Xanthine deoxyribose triphosphate is not a substrate for the in vitro DNA-synthesizing system of Kornberg. The failure of the synthesizing system to recognize xanthine as a precursor of DNA syn-

64 THE MECHANICS OF INHERITANCE

thesis suggests that it may also balk at recognizing xanthine as a bona fide component of the template.

According to our picture, nitrous acid treatment of DNA can result in the transitions AT to GC and GC to AT. A prediction of this notion is that HNO_2-induced *rII* mutants should be revertible with HNO_2. And so they are.

FIG. 4.7. *Deamination by nitrous acid (HNO_2). The final products of the deamination of cytosine and adenine are bases that have the hydrogen-bonding specificities of thymine and guanine, respectively.*

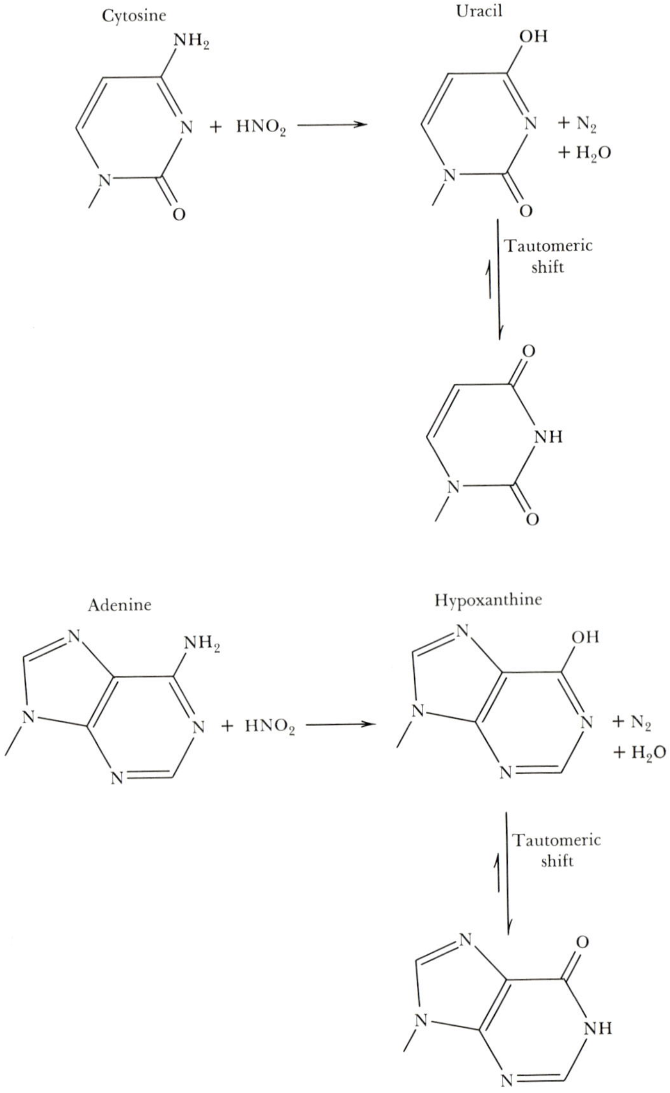

A joint consideration of our pictures of 5BU mutagenesis and HNO_2 mutagenesis leads to the prediction that these mutagens should be capable of reverting each other's induced mutants. Indeed they do. The picture is further strengthened by the reported detection of heteroduplex particles as the first products of mutagenesis by 5BU.

The reader might be able to spot a number of other predictions of the notion that 5BU and HNO_2 induce base-pair transitions. Some of these predictions have been tested; others await testing.

The notion that transition mutations arise by mistakes during duplication resulting from a failure to properly recognize a base is plausible—the experiments which support the notion make a self-consistent set. The ideas behind these in vivo experiments on mutagenesis receive further support from in vitro studies on nucleic acid synthesis (RNA in this case). Guanidylic acid (provided as guanosine triphosphate) is polymerized in the presence of polycytidylic acid (poly C) and an enzyme, RNA polymerase. If a mixture of guanosine and adenosine triphosphates is provided, the formation of only guanidylic acid is stimulated by the poly C. Thus, the poly C is acting as a template directing the synthesis of chains complementary to itself. The fidelity of the process in these in vitro conditions is attested to by the low levels at which adenine is incorporated into the poly G chains by "mistake." The mistake rate can be increased, however, by prior treatment of the poly C with either of two agents which are known to be mutagenic. Both irradiation with ultraviolet light (2,537 Å) and treatment with hydroxylamine (NH_2OH) alter the poly C templates in such a way as to increase the mistake rate as measured by the amounts of adenine incorporated into the poly G product. The hydroxylamine results correlate nicely with the observed in vivo mutagenic effects of hydroxylamine. Hydroxylamine will revert many but not all mutants induced by 5 BU or HNO_2. The revertable ones are presumably those which are GC in the mutant state and AT in the wild-type.

You should be able to devise some rigorous tests of these notions after you have read Chap. 10.

Mutational Heteroduplexes

The concept of a heteroduplex structure resulting from mutation receives support from studies on bacteriophages. If wild-type T4 particles are treated with HNO_2 (for instance) and then plated on *E. coli* strain B, a minority of plaques containing both *r* and wild-type phage may be found among the majority wild-type plaques. These mixed plaques can be spotted by their "mottled" or "sectored" appearance. Most of the *r*'s induced by the mutagenic treatment of T4 particles are recovered from such mottled plaques. Thus, a T4 particle which has responded to HNO_2 in such a way that it produces mutant offspring (that is, has been mutated by HNO_2) *retains* the ability to produce wild-type offspring. The overwhelming majority of these offspring, however, do *not* have this "schizoid" quality—a given offspring particle gives either *only* mutant or *only* wild-type progeny (except, of course, for rare spontaneous mutations which may occur when they duplicate). Our interpretation of this behavior of newly arising mutants receives support from the observed behavior of phages whose chromosome is single-chain nucleic acid. These phages do *not* give mixed plaques when plated following mutagenic treatment with HNO_2. Heteroduplexes (mottled-plaque formers) are found also following the growth of T4 in the presence of 5BU. At the level of analyses described to this point the scheme in Fig. 4.3 appears to be adequate. However, closer examination of the behavior of heteroduplexes casts some doubt on the scheme.

Segregation from Heteroduplexes

According to the duplication scheme of Watson and Crick, duplication of a heteroduplex results in the formation of two "pure" molecules, one wild type and one mutant. In phages, it is this presumed behavior of heteroduplexes which was taken as an explanation for the origin of mottled plaques. Within this scheme, one would expect a heteroduplex particle to give mutant and wild-type offspring in at least roughly equal amounts. Coinfection of a cell by one mutant and one wild-type particle does result in approximately equal numbers of the two types among the 100 to 200 particles produced. However, infection of cells by heteroduplexes resulting either from HNO_2 or NH_2OH treatment does not produce such yields. *Most* of the cells so infected give mixed yields, and the *average* yield of mutants (or wild types) is 50 percent of the total. However, the actual yields realized in individual cells are often remote from the average. Some cells produce mostly r mutant and some produce mostly wild type. In fact, any mutant frequency (from 0 to 100 percent) appears about as often as any other.

This result, which is rather unexpected within the Watson–Crick model for duplication, is a simple consequence of a specific version of Jehle's notion. Suppose that in T4 either one chain or the other of any duplex could act as the information source for both new chains and that the two choices were not only equally probable but were made afresh in each act of duplication. Such behavior would result in exactly the observed distribution of mutant and wild-type particles among the offspring of individual heteroduplexes. (Convince yourself.)

Among bacteria also the behavior of (presumed) heteroduplexes is apparently at variance with Watson–Crick predictions. In some experiments, at least, mixed populations do *not* arise from mutagenized bacteria. This result could be taken as casting doubt on the occurrence of heteroduplexes in bacteria. Or it could mean that heteroduplexes are converted into homoduplexes *before* duplication (about which we will say more in Chaps. 8 and 9). Or it could mean that Jehle's duplication model is correct and that in bacteria only a particular one of the two chains of a duplex ever acts as information source for the production of the two new chains. We need some better experiments.

Other Mutations in DNA

The mechanism of mutation proposed by Watson and Crick is the induction of base-pair transitions as a consequence of tautomerism or ionization. The validity of this hypothesis in accounting for at least *some* mutational events is substantiated by the experimental studies with HNO_2 and 5BU among others. It is clear, however, that not all mutations within DNA molecules occur by this mechanism. Some of the evidence for the existence of other kinds of mutation comes from studies of experimentally induced reverse mutation of *rII* mutants. As discussed above, transition mutations "should" be revertible by 5BU, HNO_2, or other agents that induce transitions. In keeping with this notion, it was recorded that 5BU and HNO_2 *do* generally revert their own and each other's induced mutants. Agents are known, however, that produce mutations most of which are not revertible by transition-inducing agents. Furthermore, among "spontaneous" *rII* mutants only about 15 percent can be induced to revert with transition-inducing mutagens. The nature of some of these other classes of DNA mutations and speculations as to their mode of origin are discussed below. A particularly straightforward demonstration of a "deletion" will be discussed first.

In the phage λ several mutations resulting from the loss of stretches of the DNA molecule have been characterized. These spontaneous mutants were originally detected by the perturbations they introduce in the growth cycle of this phage. One result of these perturbations is altered plaque morphology. The density of these mutant phage particles has been examined by CsCl density-gradient centrifugation and found to be less than that of wild-type λ. This reduced density of the whole particle suggests that the ratio of DNA (density in CsCl = 1.7 gm/cc) to protein (density in CsCl = 1.3 gm/cc) is less in these mutants than in wild-type λ. If it is assumed that this change in ratio is a result solely of a reduction in the amount of DNA, it can be calculated for one of these mutants that 15 percent of the nucleotide pairs are missing. The validity of this estimate is verifiable by independent methods. Direct chemical analysis might well work, but, in fact, a somewhat easier method was employed.

Phage particles containing large amounts of radioactive phosphorus (P^{32}) in their DNA are measurably unstable; they are rendered incapable of forming plaques (they "suicide") as a result of the disintegration of the incorporated P^{32}. Decay of a P^{32} atom results in rupture at the level of the decay of the DNA chain in which that atom was incorporated. With a probability of about 0.1, a decay leads to the nearby simultaneous scission of the other chain of the DNA molecule. Such a double-chain scission results in inactivation of the particle. The rate at which a population of P^{32}-labeled particles commits suicide depends on the rate at which lethal decays occur within the particle; this in turn depends on the number of P^{32} atoms in the particles. For two populations containing the same ratio of P^{32} to total phosphorus atoms, the one with particles containing the larger amount of DNA will contain the larger number of P^{32} atoms and will, consequently, suicide faster. Wild-type λ and the λ mutant with the density reduction corresponding to a 15 percent reduction in DNA content were grown to give the same extent of labeling per P atom; the mutant was observed to suicide 15 percent slower than wild type. Since wild-type λ contains about 70,000 nucleotide pairs, the mutant appears to be missing about 10,000 nucleotide pairs. Other, more conventional methods for estimating molecular weights of DNA give results which support this conclusion. Seldom are mutations that are known to be within the limits of a single DNA molecule sufficiently gross to submit to such straightforward characterization. Mutations to *rII* resulting from deletions of smaller stretches of DNA in the *rII* region of T4 have been well studied, however. A description of the method of analysis of these deletions is best deferred until Chap. 7, but the nature of some of the conclusions may be indicated here.

Deletions occur ranging in size from 1 nucleotide pair up to as much as or more than 2,000 nucleotide pairs. Only the very smallest of the deletions revert to wild type at a measurable rate (mutation rate greater than about 10^{-9}). The usual failure of deletions to revert is easily understood; gross changes in the nucleotide sequence of a DNA molecule are not likely to be repaired by the chance process of mutation.

The rate of appearance of small deletions or additions can be experimentally increased. In Chap. 10 we shall discuss the evidence that the organic compounds called acridines induce deletions *and* additions of small numbers (one, two, a few) of base pairs. Nitrous acid applied to free phage particles induces deletions also, although often of much greater extent. In contrast to the case for transition induction, I know of no simple sypothesis for the mechanism of deletion induction by HNO_2.

Mutagenic Mutations

Mutants of both phage and bacteria have been found which produce an altered enzyme component of the DNA-duplicating machinery. Some of these mutants have high mutation rates! In some cases the rate of occurrence of only one particular class of mutations (for example, transversions) is exalted. The enzymatic details of DNA duplication may well be revealed by future studies of these mutants.

Summary

A variety of kinds of mutations (heritable alterations) have been shown to occur within the limits of single DNA molecules. Some of these involve the substitution of one base pair by another (transitions and transversions) while others involve small additions or deletions. Large deletions occur also. We have focused our discussion on those mutations for which there exists a promising explanation in terms of nucleic acid structure. Genetic systems of higher order of complexity than those of viruses and bacteria may undergo other kinds of mutations. A comprehensive discussion of such mutations can be found in *Cytogenetics,* by Swanson, Merz, and Young, in this series.

Mutations can originate either as a mistake during duplication or as a result of chemical alteration of the parent nucleic acid molecule prior to duplication. A heteroduplex is often the primary product in either case. The analysis of the kinds and frequencies of offspring produced by duplication of heteroduplexes should make it possible to decide between competing hypothesis regarding the duplication of DNA.

References

Freese, E., "The Specific Mutagenic Effect of Base Analogues on Phage T4," *J. Mol. Biol., 1* (1959), 87–105. The Watson–Crick hypothesis for the origin of mutations begins to prove fruitful.

Kreig, David R., "Specificity of Chemical Mutagenesis," *Progress in Nucleic Acid Research, 2* (1963), 125–68. Rarely has so much been said so well in so few words.

Luria, S. E., "The Frequency Distribution of Spontaneous Bacteriophage Mutants as Evidence for the Exponential Rate of Phage Reproduction," *Cold Spring Harbor Symp. Quant. Biol., 16* (1951), 463–70. Reprinted in *Papers on Bacterial Viruses,* G. S. Stent, ed. (Boston: Little, Brown & Co., 1960), pp. 139–50. Viruses became organisms as a consequence of Luria's presentation—mutationally speaking, they get and beget like the rest of us.

Speyer, J. F., J. D. Karam, and A. B. Lenny, "On the Role of DNA Polymerase in Base Selection," *Cold Spring Harbor Symp. Quant. Biol., 31* (1966), 693–97. Phages that make mutant DNA polymerase have a high mutation rate.

Terzaghi, B. E., George Streisinger, and F. W. Stahl, "The Mechanism of 5-Bromouracil Mutagenesis in the Bacteriophage T4," *Proc. Natl. Acad. Sci. U. S., 48* (1962), 1519–24. Some 5BU-revertible mutations in phage T4 can revert only when 5BU is present in the intracellular environment of the multiplying phage; others can revert when 5BU-substituted particles are permitted to duplicate in an ordinary intracellular environment.

70 THE MECHANICS OF INHERITANCE

Watson, James D., and Francis H. C. Crick, "The Structure of DNA," *Cold Spring Harbor Symp. Quant. Biol., 18* (1953), 123–31. Reprinted in *Papers on Bacterial Viruses,* 2nd ed., G. S. Stent, ed. (Boston: Little, Brown & Co., 1965), pp. 230–45. A discussion of the evidence underlying the Watson–Crick formulation for the structure of DNA and their hypothesis for the mechanisms of duplication and mutation that derive from it.

Problems

4.1. Some bacteria were infected with exactly one each of wild-type T4. The yields from many individual cells ("single bursts") were then examined for mutations to the *r* genotype. The total yield from each cell was about 200. Of 10,000 single bursts examined, 200 had at least one *r* mutant. What is the mutation rate to *r*?

4.2. Consider a heteroduplex DNA molecule whose base sequence is shown in part below

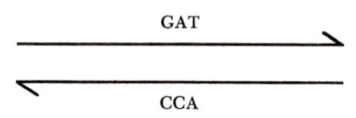

```
         GAT
  ─────────────────────▶

  ◀─────────────────────
         CCA
```

Suppose this duplex duplicated for three generations (giving eight molecules):

(*a*) If duplication were according to the scheme of Watson and Crick, what fraction of the eight molecules would be heteroduplexes?
(*b*) If duplication were according to Jehle, what fraction of the eight molecules would be heteroduplexes?
(*c*) If duplication were according to Jehle with the further specification that at each duplication the chain to act as information source for the two new chains is chosen at random, what is the probability that a molecule picked at random from among the eight is of composition shown below?

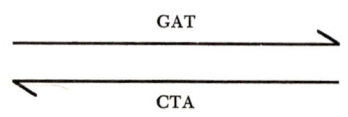

```
         GAT
  ─────────────────────▶

  ◀─────────────────────
         CTA
```

(*d*) On the Watson–Crick scheme what would be the probability of obtaining the molecule in (*c*)?

4.3. (*a*) Phage particles treated with nitrous acid lose their ability to form plaques; with increasing dose, a decreasing fraction of the particles make plaques. The following sorts of data can be obtained:

Duration of treatment with constant HNO_2 concentration (min)	Fraction of phage surviving
0	1
1	0.1
2	0.01
3	0.001

How many deaminations must occur to kill a particle?

(b) Examination of the survivors of HNO_2 treatment reveals that the frequency of mutants among survivors is increased with increasing dose. Data for the rate of reverse mutation of any particular rII marker are easily obtained by plating an HNO_2-treated rII stock on *E. coli* K(λ), on which only r^+ and r^+/rII heteroduplex phages make plaques. The primary measurement made is (live r^+ and r^+/rII) per milliliter. Suppose 10^9 phages per milliliter were treated under the same conditions used in Problem 4.3a, and the following data were obtained:

Duration of treatment (min)	(Live r^+ and r^+/rII) per milliliter
0	10
0.05	460
0.1	810
0.5	1,600
1	1,000
2	200
3	30

Assume that killing and mutagenesis are independent events. (1) How many deaminations are required to make a phage that will plate on K(λ)? (2) What fraction of the live phages are induced to revert per minute of treatment?

(c) At any dose, what is the ratio of reversions at the marker studied to all lethal deaminations?

4.4. Particles of a strain of phage suspected of "carrying" a genic deletion were grown along with wild-type λ in an environment heavily laced with P^{32}. The P^{32}-labeled offspring particles were collected and the titers of viable particles of each type were determined at intervals. Suppose the following data were obtained:

Fraction of P^{32} decayed	Titer of wild type	Titer of deletion mutant
0	1.0×10^9	1.0×10^9
0.2	9.5×10^7	1.2×10^8
0.4	9.0×10^6	1.5×10^7
0.6	8.5×10^5	1.7×10^6

What fraction of the wild-type DNA is missing in the deletion mutant?

Five

Organization of Genic Material

The purpose of this chapter is threefold: (1) to examine the spatial organization of DNA within a creature; (2) to examine the way in which DNA organization is modified in time as a creature proceeds through its life cycle; and (3) to make these examinations for creatures of varying degrees of biological complexity. We can recognize three levels of complexity without making any particular effort to justify the selection beyond the recognition of their heuristic convenience.

Level I includes the viruses. We may characterize the creatures of this level as those that can multiply only *within* cells because of their own lack of essential energy-providing and synthetic metabolic processes. They have nucleic acid associated (in some cases) at some stages in their life cycle with protein that aids in their dissemination.

At level II are those creatures that *can* lead relatively independent existences (some are even photosynthetic!) but that lack important structural features characteristic of the cells of creatures in level III. Another name for the numerous organisms that may be grouped in level II is bacteria.

In level III are all those creatures composed of various and varying numbers of cells each of which has a nucleus clearly differentiated from a cytoplasm. Most of the DNA of such creatures resides in the nuclei of their cells where

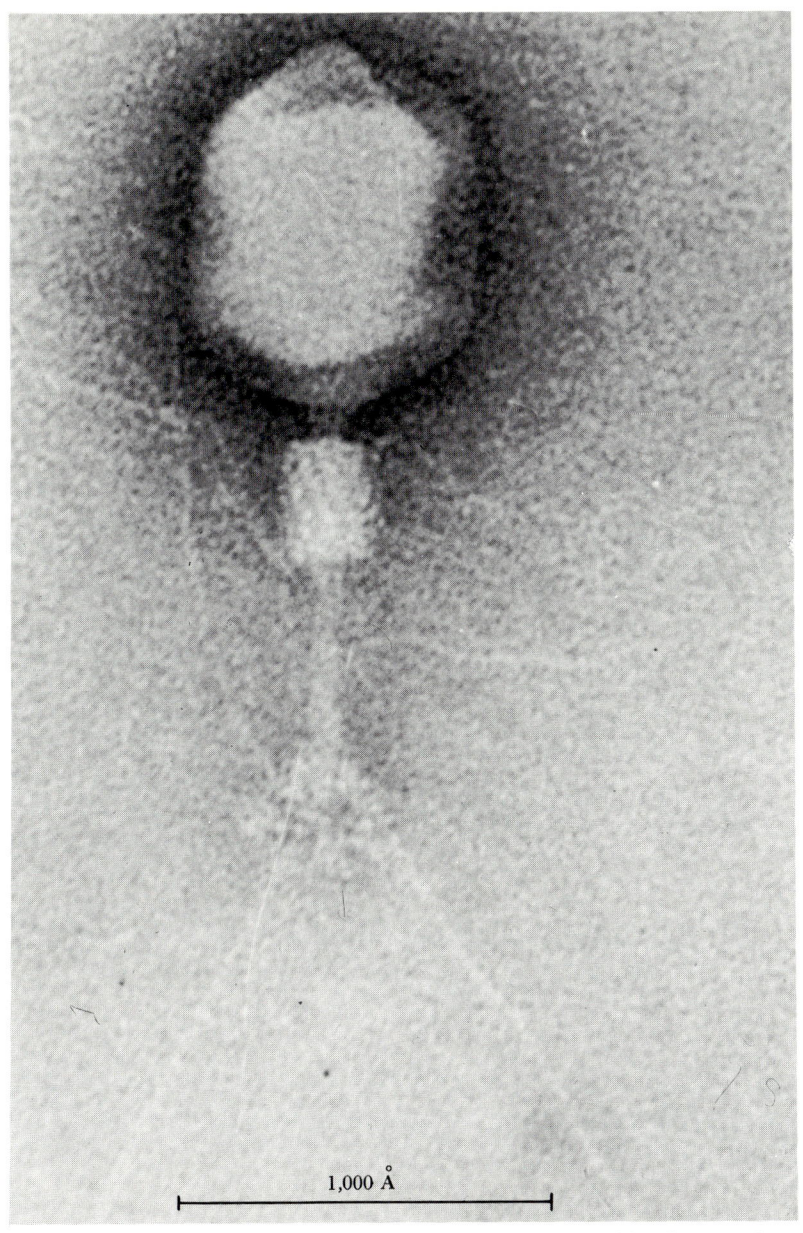

FIG. 5.1. *An electron micrograph of a T4-phage particle. This photograph was selected because the method by which it was prepared (phosphotungstic acid embedding of the specimen) accurately preserves the dimensions of the specimen. The appearance of the tail of the particle is abnormal for reasons that do not interest us now. The volume into which the DNA is packed within the 50 Å thick protein "skull" can be estimated from this figure to be 2×10^{-16} cm³. This number is about equal to the volume of the DNA of this phage calculated from the Watson–Crick dimensional parameters for DNA. This photograph, kindly supplied by R. W. Horne, appeared in* J. Mol. Biol., 1 *(1959), 281.*

74 THE MECHANICS OF INHERITANCE

it is organized into Chromosomes.[1] For convenience, the creatures in the three levels will be called viruses, bacteria, and higher organisms respectively.

Packing of DNA

Creatures at all levels face a common intriguing problem relative to the organization of their DNA. For each of them, there is a stage (or there are stages) in the life cycle when the DNA is very tightly packed into a small space. Our notions about DNA organization must take this property of genic structures into account while admitting that its basis is not understood (Fig. 5.2).

We may illustrate the "packing problem" by considering the dimensions of a mature, infectious T4 particle and of the DNA packed within it.

The T4 DNA has a total volume that we can calculate using the dimensional parameters for DNA given in Fig. 2.14 in conjunction with the chemically determined nucleotide content of a T4 particle of 2×10^5 nucleotide pairs. We may approximate the shape of a DNA molecule as a cylinder. Then:

$$\text{Volume of T4 DNA} = \text{cross sectional area of DNA} \times \text{total length of T4 DNA}$$

$$= \pi (10^{-7} \text{ cm})^2 \times \left(\frac{3.4 \times 10^{-8} \text{ cm}}{\text{nuc. pair}}\right)(2 \times 10^5 \text{ nuc. pairs})$$

$$= 2 \times 10^{-16} \text{ cm}^3$$

The volume into which this DNA is packed can be estimated from electron micrographs of mature phage particles like the one in Fig. 5.1. With the aid of the scale given in that figure you can calculate this volume. The figure (not including the tail) is a bipyramidal hexagonical pyramid. An outer layer of protein of about 50 Å thickness surrounds the DNA; lateral dimensions taken from the figure should be appropriately reduced to be applicable to the DNA. Using the micrograph in Fig. 5.1, I calculated 2×10^{-16} cm³ as the volume into which the phage DNA is packed.

The nature of the packing problem is more dramatically illustrated by a comparison of the *length* of the phage DNA with the *diameter* of the phage head. If we approximate the morphology of the phage

---

[1] *Chromosome* with a capital *C* will be used to refer to the genic structures of higher organisms; the word was originally coined in their behalf. It is capitalized out of respect and also to distinguish it from *chromosome* with a lower case *c,* which will refer to the genic structure of *any* creature.

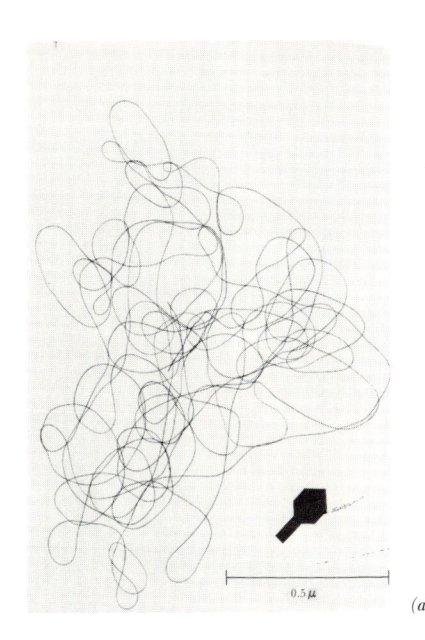

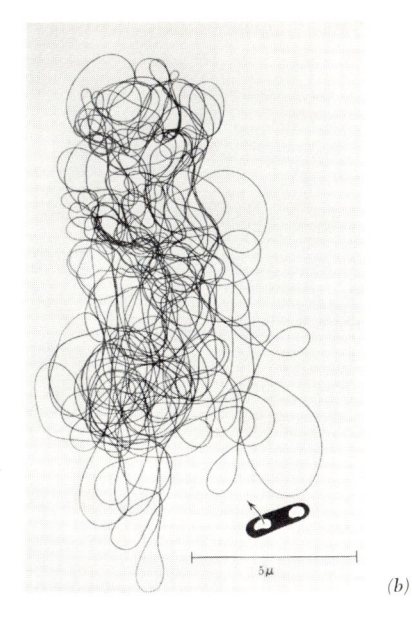

FIG. 5.2. *The total length of the DNA of a virus, a bacterium, and a higher organism compared to scale with the structures into which the DNA is organized. The three parts of the figure are not to the same scale. (a) Bacteriophage T4. The chromosome of T4 is about 60μ long. (b) Bacterium E. coli. The chromosome of E. coli is about $1,000\mu$ long (1 mm). (c) Drosophila melanogaster, a higher organism. The total DNA complement of a cell of a fruit fly is about 16 mm long. The figure depicts the Chromosomes as they appear in mitotic metaphase (see page 85) in a cell containing two complete sets of Chromosomes (a diploid cell).*

head by a hollow sphere of comparable internal volume, it would have an inside diameter of a bit less than 800 Å. This dimension is about 1,000 times smaller than the total length of the DNA in T4 (see Fig. 5.2a).

As indicated previously, the packing problem is not unique to viruses. In Figs. 5.2b and 5.2c the total length of DNA in one genome (one complete set of genic material) for bacteria and for the fruit fly *Drosophila* are compared to scale with the structures into which that DNA is packed at some stages of the life cycles of the respective cell types.

Virus Chromosomes

When viruses are gently disrupted the entire nucleic acid (DNA or RNA) content of each particle can be isolated as a single nucleic acid molecule (the chromosome of the virus; see Frontispiece and Fig. 5.3). Among the larger viruses the chromosome is duplex (Watson–Crick) DNA and has two (only) ends. Among the smaller viruses, however, RNA chromosomes are common. Furthermore, some of these small chromosomes are single-chained, others double-chained. Cutting across the categories of RNA versus DNA and single- versus double-chain chromosomes are the categories of linear (two-end) versus circular (no-end) chromosomes.[2] The following criteria are among those used for categorizing virus chromosomes according to the scheme described above: (1) The presence of ribose versus deoxyribose defines the RNA or DNA nature of the chromosome. (2) The lack of equality in the molar amounts of adenine and thymine (or uracil) or guanine and cytosine signals a single-chain chromosome. Electron microscopy and analysis of hydrodynamic properties of the chromosomes help too. (3) Electron microscopy is the most convincing technique for distinguishing between circular and linear chromosomes.

Other variations in virus chromosome structure occur. The chromosomes of some duplex DNA viruses have occasional interruptions in the sugar-phosphate backbones, like this:

Chain interruption

In those cases where both chains bear interruptions, interruptions on opposite chains are sufficiently far apart that the forces stabilizing the duplex (hydrogen-bonding and base-stacking) hold the chromosome

---

[2] However, no circular RNA chromosomes have been reported to date.

Organization of Genic Material 77

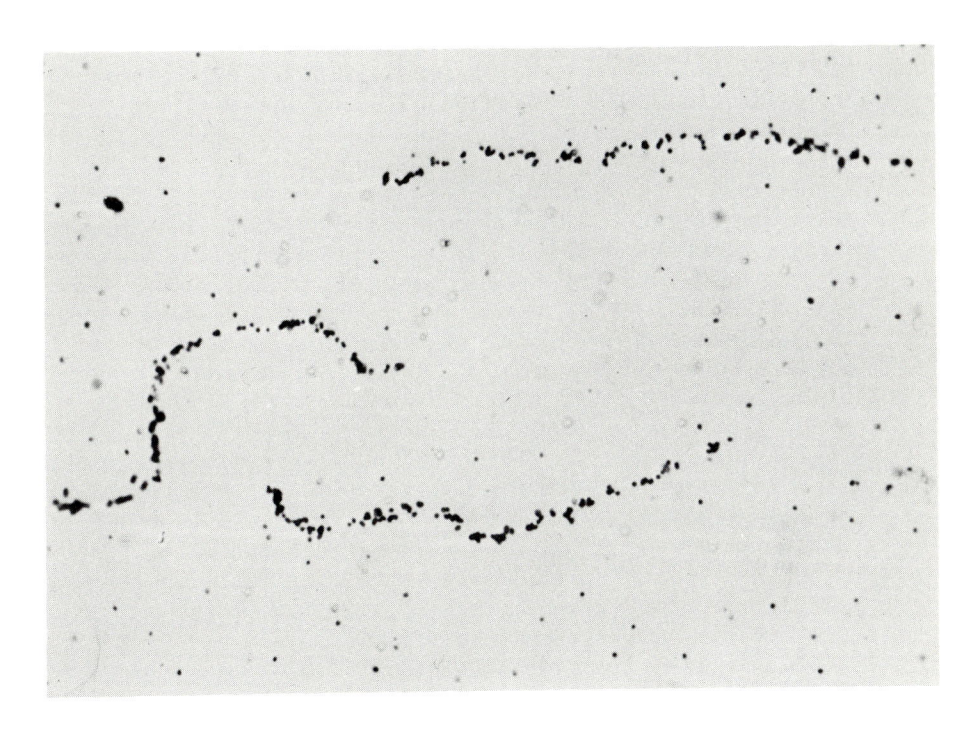

FIG. 5.3. *Autoradiographic images of isolated chromosomes of the phage T4. Bacteria growing in medium containing tritium (H³)-labeled thymidine were infected by phages. The progeny phages emerging were isolated and gently disrupted. The disrupted particles were suspended in melted photographic emulsion that was then gently poured onto a smooth glass plate. When the emulsion had hardened, the plates were stored until an appreciable number of H³ atoms in each chromosome had decayed. The electrons ejected from the nucleus of a decaying H³ atom travel but a short distance through the emulsion, sensitizing a small number of silver grains. Upon development of the emulsion, the sensitized grains appear as very short "tracks." The length of each phage chromosome can then be measured by measuring the length of the photographic image composed of a row of a number of short tracks. In this photograph the chromosomes of T4 appear to be about 50 μ long. Sixty-eight μ is the length of the total amount of DNA per phage calculated from the chemically measured nucleotide content of T4 and the dimensional parameters of the Watson–Crick DNA model. This photograph was kindly supplied by John Cairns.*

together. Some viruses have chain interruptions at special places; in other viruses the interruptions appear randomly disposed. The chromosomes of some viruses are usually free of chain breaks. Maybe a future edition of this book can assess the significance of these structural variations; suggestions are invited. In Chap. 7 yet other features of the structure of chromosomes of virus particles are revealed.

In the course of the T4 life cycle, the DNA undergoes transitions from the condensed phase to the extended phase and vice versa. In the mature particle the DNA is condensed; throughout the intracellular phase extended DNA is present. Indeed, it is only in the extended phase that the DNA directs protein synthesis, duplicates, or undergoes genetic recombination (see Chapt. 7). About halfway between the time of infection and the time of cellular lysis, mature particles containing condensed DNA appear in the infected cell.

Little is known about the mechanism of the transition of phage DNA from the extended form to the condensed form or its reverse. Careful electron microscopic examination of the components of mature particles and of the stages in phage maturation, however, adds detail to our knowledge of the sequence of events and permits educated speculation about some of the mechanisms.

At the time of infection the core in the phage tail probably penetrates the cell surface (see Fig. 2.5). It seems likely that the hole through the axis of the core then forms a canal between the interior of the phage head and the interior of the bacterial cell. The DNA might then "diffuse" from the region of high concentration through the canal into the bacterium. The awkward point in this scheme is getting one end of the chromosome threaded into the very narrow (20 to 30 Å diameter) hole. One may imagine that this was achieved as one of the steps in maturation of the particle in its previous host.

As might be anticipated, condensation of the DNA does not represent a reversal of the above steps; the DNA is not "stuffed" into otherwise completed T4 particles. Instead, it appears as though the tails and tail fibers are attached to heads already filled with DNA. The sequence of steps in the construction of filled heads is unknown. The central question is whether head protein is built around condensed DNA or whether the DNA is condensed into previously built heads.

The Bacterial Chromosome

Our knowledge of the structure of viral chromosomes is in pretty good shape. For bacterial chromosomes the problem is more difficult because they are bigger. We may summarize our state of knowledge by stating that none of the physical methods brought to bear has indicated a structure any more complex than a single (very long) duplex DNA molecule. Electron micrographs like that in Fig. 5.4 constitute one kind of evidence supporting that view. The analysis of bacterial conjugation conducted by F. Jacob and E. L. Wollman is another.

Conjugation in *E. coli* involves the formation of a conjugation tube between members of opposite sex. DNA from one of the nuclei of the

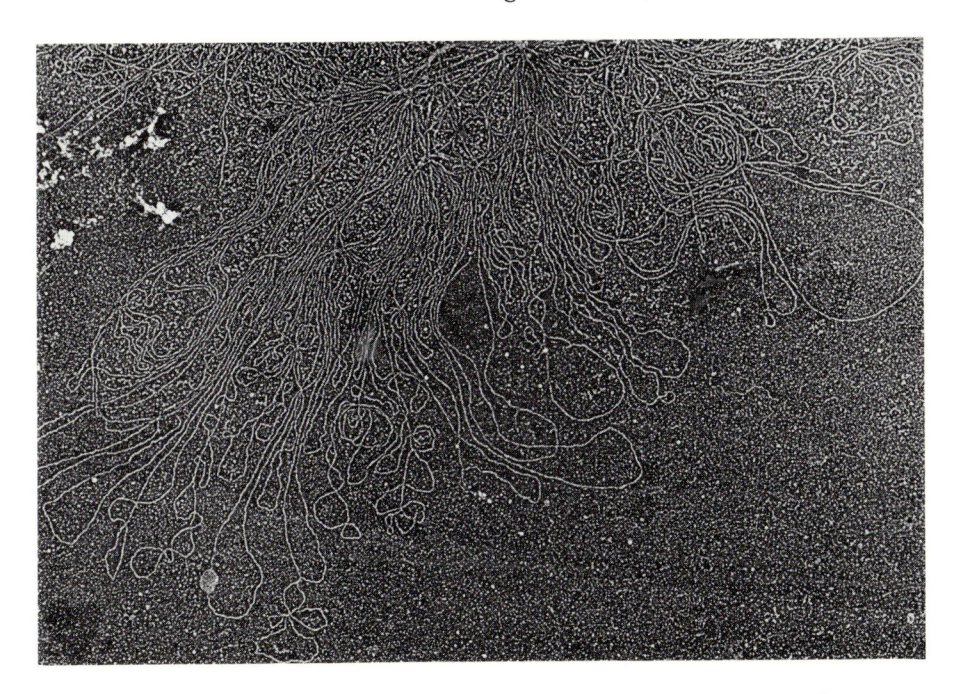

FIG. 5.4. *An electron micrograph of part of the DNA from a ruptured cell of the bacterium* M. lysodeiktikus. *The rarity of "ends" in the photograph suggests that the DNA is organized in one long unbranched structure. This photograph, kindly supplied by A. K. Kleinschmidt, appeared in* Z. für Naturforschung, 16b *(1961), 730.*

"male" cell passes slowly through the tube into the "female" cell. Conjugation is completed upon rupture of the tube, which may occur either before or after all the DNA has been transferred. After a short lag, division of the female cell resumes. If the two cells undergoing conjugation differ by one or more distinguishable hereditary characteristics, the DNA transfer may have observable genetic consequences; descendents of the female exconjugant may manifest some of the transferred genetic markers. In Chap. 8 we shall examine in more detail the fate of the transferred DNA. For our present purposes, it is sufficient to note that transfer can be detected by its genetic consequences. The experiment described below suggests that the genetic markers are located on a single linear structure. The results are fully compatible with the picture of the bacterial chromosome gained by direct observation.

Conjugation in bacteria can be interrupted at will by subjecting a suspension of conjugating cells to the shear forces created by rapid stirring. For any conjugating pair, the interruption of conjugation leads to the failure of some markers to be transferred; apparently the

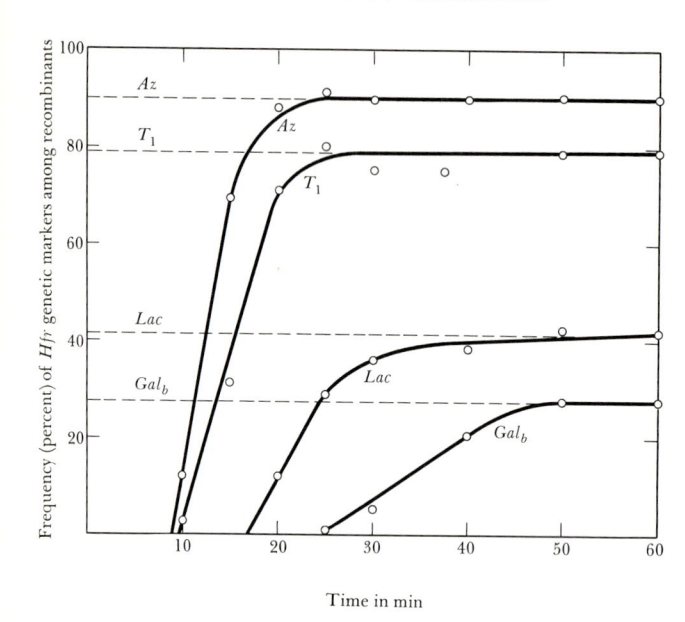

FIG. 5.5. *The order of penetration of the genetic characters of the male strain K12 Hfr H of E. coli during conjugation. Male and female (K12 F-) bacteria were mixed together under conditions permitting conjugation to occur. At intervals samples were removed and conjugation was interrupted by treatment with a blender. The samples were then spread on an agar-petri plate under conditions that permit the growth of certain descendants of the female exconjugants only. In order to grow, a cell must have one distinguishing hereditary characteristic of the F- strain (resistance to streptomycin in this case); this ensures that it be a descendant of a female cell. In addition, the cell must possess one (or more) hereditary characteristics of the Hfr strain; this ensures that the cell is indeed a descendant of an exconjugant. The hereditary abilities to make the amino acids threonine and leucine (T+L+) were selected for in this case. These two markers are known to be transferred very early (before 10 min) in this Hfr strain. The time-axis intercepts of the curves represent the times at which each of four other genetic markers are first transferred into some of the F- cells. The final values achieved by the curves are determined in part by the likelihood of spontaneous interruption of conjugation and in part by events that occur after tranfer (genetic recombination; see Chapt. 8) and that can lead to the failure of transferred markers to be transmitted into the same daughter cells of the F- exconjugant. From François Jacob and E. L. Wollman,* Sexuality and the Genetics of Bacteria, *rev. ed. (New York: Academic Press, Inc., 1961).*

chromosome is broken along with the conjugation tube. In a population in which the onset of conjugation has been synchronized, the time at which each of the markers is transferred can be determined by artificially interrupting conjugation at various times, and scoring the female exconjugants for transferred markers. For a given male strain it is found that each of the many genetic markers studied has a unique time of transfer. Data and further details of this experiment are given in Fig. 5.5. In Chap. 8 we shall examine other aspects of bacterial conjugation.

The outcome of another kind of experiment is consistent with our view of phage and bacterial chromosomes as being single DNA molecules or, perhaps, an array of DNA molecules attached end to end. This experiment, which may take a variety of forms operationally, involves an examination of the distribution of labeled parental atoms between daughter nuclei in the case of bacteria or among progeny particles in the case of phages.

For phages, equilibrium density-gradient centrifugation of intact particles has been applied. The large percentage of DNA in phage particles and the stability of the particles in CsCl are responsible for the success of the approach. In the case of bacteria, autoradiography of daughter cells of uninucleate bacteria labeled in their DNA with H³-thymidine has been successfully applied. When complications resulting from genetic recombination (see Chaps. 7 and 8) are accounted for or avoided, the genomes of both phages and bacteria are observed to be composed of two subunits that separate from each other at duplication and remain intact throughout subsequent generations; that is, the chromosomes of both phages and bacteria act with respect to the distribution of atoms at duplication like single DNA molecules.

Chromosomes of Higher Organisms

In bacteria the separation of daughter chromosomes from each other prior to cell division can be observed microscopically (Fig. 5.6). It is obvious that for bacteria such direct visual observation is of little help in elucidating structure and behavior of chromosomes at the molecular level. The situation is somewhat better for many higher organisms. The Chromosomes, though more complicated in structure, are at least large enough to see clearly. Other volumes in this series describe aspects of Chromosome appearance and behavior which hold important positions in our understanding of heredity. This volume will restrict itself to the few observations which seem to bear most directly on the structure and mode of duplication of Chromosomes.

The Chromosomes are composed of DNA, RNA, and two classes of proteins, in addition to smaller amounts of other things. Our picture of the roles played by each of the major components is based primarily on inference. From the studies on microorganisms we infer that the role of DNA in Chromosomes is to carry genetic information from generation to generation. Our present knowledge of how the information in DNA is expressed (see Chap. 10) permits us to identify a role of the RNA. The RNA may be thought of as a family of polymers made on DNA templates and destined primarily for transfer to the cytoplasm where they are involved in the synthesis of proteins. There is no evidence supporting the possibility that any of the RNA of higher forms serves as genic material as it demonstrably does for some viruses. Our compulsion for simplicity leads us to ignore this possibility from here on. Of the two classes of proteins, one (the histones) contains many basic amino acids (amino acids with extra NH_2 groups) and little tryptophan. These proteins are bound to the DNA by ionic bonds.

82   THE MECHANICS OF INHERITANCE

It is redundant to say that they serve to neutralize the phosphate groups of the DNA, which, at physiologic pH, are highly ionized, but at present no other role is known for them. The spatial relationship between DNA and basic protein as deduced from X-ray diffraction studies is shown in Fig. 5.7. The other, nonbasic proteins are characterized by higher tryptophan content as well as by lower content of basic amino acids.

Each time a cell divides to become two cells, the two daughter

FIG. 5.6. *Separation of daughter chromosomes in* E. coli *as viewed by time-lapse photomicrography. A small clump of cells growing with a generation time of 30 min was photographed at intervals under conditions of illumination that distinctly reveal the chromosomes (phase contrast microscopy). The array of photographs should be read "like a book," from left to right starting on the top line. The individual photographs were taken at the times (in minutes) indicated below (letting 0 be the time of the first frame):*

0	6	12	21	30
33	36	39	48	54
57	60	66	72	75

These photographs were kindly supplied by D. J. Mason *of the Upjohn Company, Kalamazoo, Michigan.*

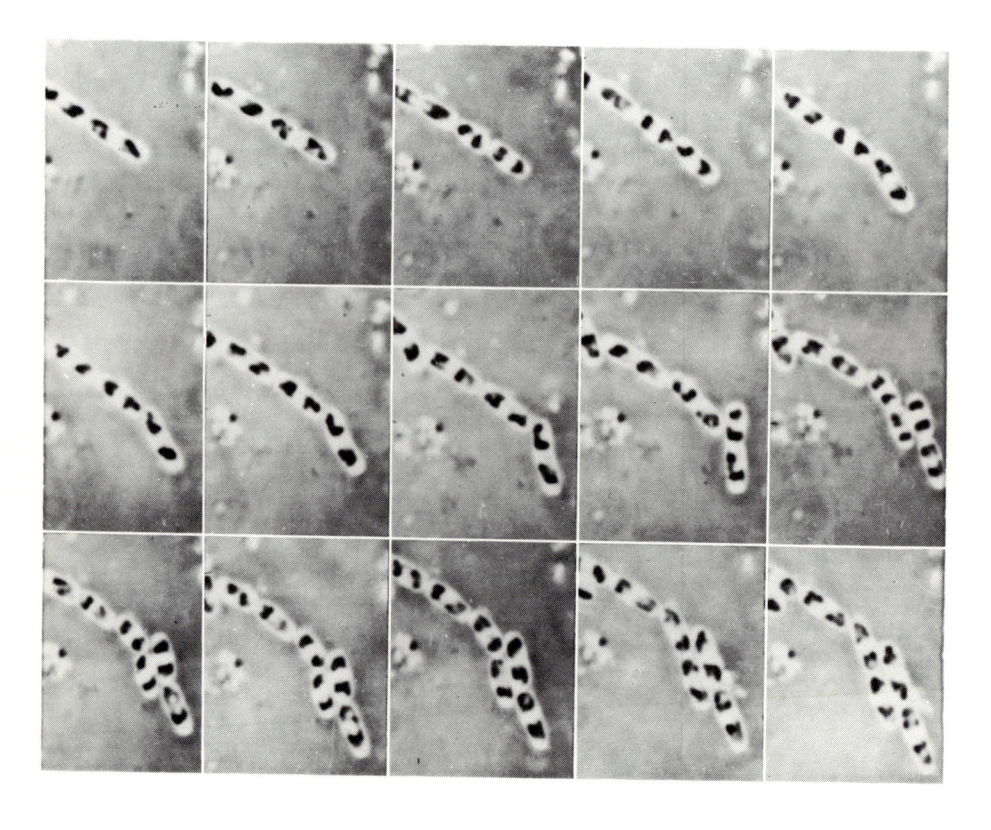

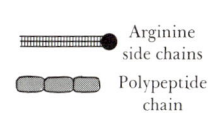

FIG. 5.7. *Diagram showing the relationship between DNA and the basic protein protamine suggested by X-ray diffraction data. The small groove of the DNA molecule is occupied by the polypeptide chains that have a large number of basic residues. The amino groups of these residues lie close to the phosphate groups of the two DNA chains. Nonbasic amino acids in the protamine are accommodated by folds in the peptide chains. Protamines occur exclusively in the Chromosomes of the sperm of some species. The more common basic protein, histone, neutralizes the charged phosphate groups in the DNA of most other Chromosomes. X-ray diffraction data suggest that the histones occupy both grooves of each DNA molecule. This figure, reproduced by permission of Maurice Wilkins, appeared in* Cold Spring Harbor Symp. Quant. Biol., 21 *(1956), 83.*

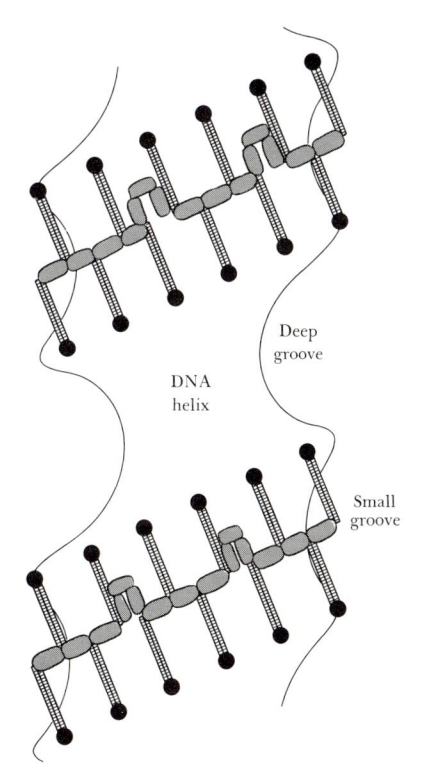

cells are essentially identical to the mother cell; in particular, each daughter receives a full complement of Chromosomes. Two activities of the Chromosomes are thereby implied. (1) Chromosomes duplicate between cell divisions. (2) Daughter Chromosomes are parceled out to daughter cells so that each cell receives one full set of Chromosomes.

During most of the division cycle, Chromosomes are not visible as discrete entities; their DNA is in the elongated phase (Fig. 5.8a). During this period the DNA duplicates. The duplication can be detected, among other ways, by autoradiographically determining the time of incorporation of H^3-labeled thymidine. Subsequent to the completion of DNA duplication, condensation of the Chromosomes begins (Fig. 5.8b). As the Chromosomes become visible they can be seen to divide longitudinally; at this stage each Chromosome is composed of two chromatids. When condensation is well advanced, the nuclear membrane disintegrates and the spindle apparatus appears (Fig. 5.8c). The Chromosomes assume positions on the equatorial plane of the spindle axis (Fig. 5.8d). Spindle fibers appear to connect the centromere of each Chromosome to each of the two poles of the spindle. The centromeres become visibly double (Fig. 5.8e), and the

84   THE MECHANICS OF INHERITANCE

FIG. 5.8. *A diagrammatic summary of the behavior of Chromosomes during mitosis. For clarity of illustration (both visual and conceptual) I have "selected" a haploid cell containing a complement of only two nonhomologous chromosomes. This explanation is not logically necessary—I offer it to help the reader who may have encountered mitosis previously; the naïve reader is asked to ignore it.*

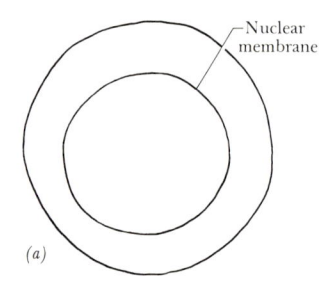

(a) *During interphase, Chromosomes are typically so elongated as to be invisible to ordinary microscopic examination. DNA duplication occurs during this period.*

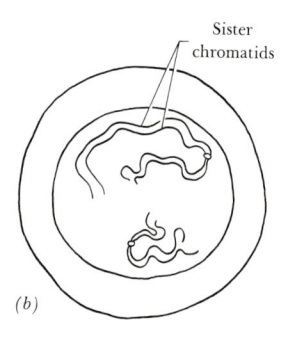

(b) *The onset of prophase is signaled by a degree of Chromosomal condensation (shortening and thickening) that makes the Chromosomes visible. At this stage, each Chromosome is visibly composed of two longitudinal halves, the sister chromatids.*

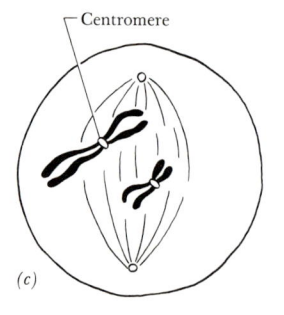

(c) *The spindle apparatus, which is responsible for the orderly separation of the sister chromatids, appears as the nuclear membrane disappears and the Chromosomes continue to condense.*

(d) *In metaphase, the Chromosomes assume positions on the equatorial plane of the three-dimensional spindle apparatus. The centromeres appear to be connected to the two spindle poles by fibers of the spindle apparatus.*

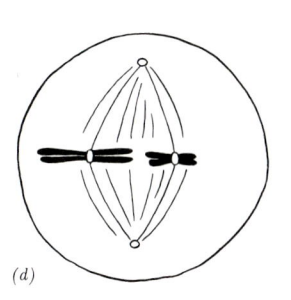

(e) *The centromeres of each Chromosome divide.*

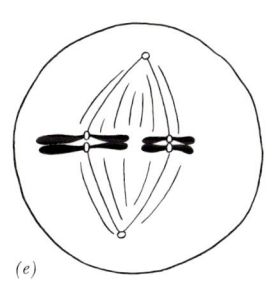

(f) *In anaphase the daughter Chromosomes move to opposite poles, aided in their movement by the spindle apparatus.*

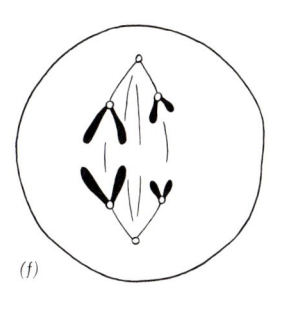

(g) *The mitotic cycle is completed by reentry into interphase. The spindle fibers disappear, the nuclear membrane reforms, and the chromosomes elongate. The early stages in this portion of the cycle are referred to as telophase. In most cases, cytoplasmic division follows nuclear division in such a way that the two cells each receive a nucleus containing identical sets of chromosomes.*

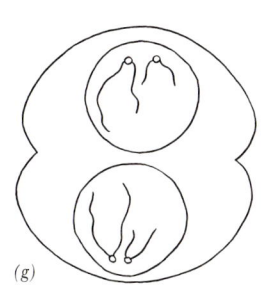

two halves of each Chromosome move to the opposite poles of the spindle (Fig. 5.8f). The nuclear membranes reform, the spindle apparatus disappears, and the Chromosomes return to the elongated state (Fig. 5.8g). Nuclear division is followed by cytoplasmic division and the complete return of the nucleus to the interphase state. The conventionally identified "stages" in this process called mitosis are diagrammed, described, and named in Fig. 5.8.

The preceding description of mitosis is both simplified and generalized. It will suffice for our purposes, however.

J. H. Taylor and his associates studied the transfer of atoms of the DNA of Chromosomes to their daughter Chromosomes. DNA in the root tip of English broad bean (*Vicia faba*) seedlings was labeled with H^3 by exposing the growing root tip to a solution containing H^3-labeled thymidine. After about a third of a division cycle, the seedlings were transferred to a nonlabeling medium containing colchicine. The colchicine permits Chromosome duplication while blocking cell division. After periods corresponding to one or two cell-generation times, seedlings were removed and the root tips fixed and pressed against photographic film. The electrons emitted upon the decay of the incorporated tritium are of such low energy that they produce images only immediately above their point of entry into the film. The level of labeling in each of the two chromatids of a metaphase Chromosome can therefore be separately determined. The film is developed while opposed to the pressed root tip; the Chromosomes and their autoradiographic images can be simultaneously viewed. The number of divisions undergone by a set of Chromosomes since the time of labeling can be determined by counting the Chromosomes. The Chromosomes in cells with 12 metaphase Chromosomes have not duplicated following the labeling. The Chromosomes in cells with 24 and 48 metaphase Chromosomes have duplicated once and twice respectively.

The Chromosomes in labeled nuclei undergoing their first metaphase since the period of labeling were equally radioactive in the two chromatids. The 24 Chromosomes in nuclei in their second postlabeling metaphase were radioactive in one chromatid only; the amount of label in the radioactive chromatid was equal to the amount present in each chromatid in the previous metaphase. In those cells with 48 Chromosomes, two of the sets of Chromosomes were completely unlabeled; the other two sets were labeled in the same fashion as the Chromosomes in cells that had undergone one less division. The reader should convince himself of the validity of Taylor's conclusion: "The chromosomes before duplication are composed of two units which extend throughout the length of the chromosome. The units separate at duplication and each has a complementary [3] unit built along it." In short, Chromosome

---

[3] The complementary character of the two Chromosome units is deduced from the rules that apparently govern the (rare) "sister-strand" exchanges observed in these experiments (see Problem 5.3).

Organization of Genic Material   87

FIG. 5.9. *The mode of distribution of the atoms of parental DNA into sister chromatids as revealed by the experiments of J. Herbert Taylor and his collaborators.*

(a) An interphase Chromosome is composed of two subunits. DNA duplication is permitted to proceed in the presence of radioactive thymidine (†). Following DNA duplication, four subunits are present. You can't see the chromatids in interphase.

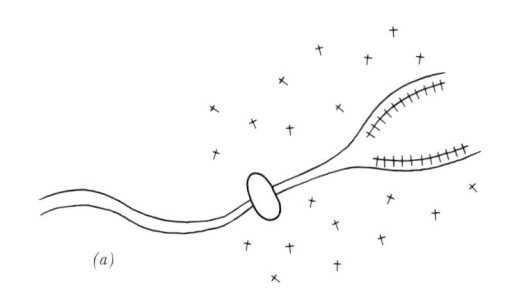

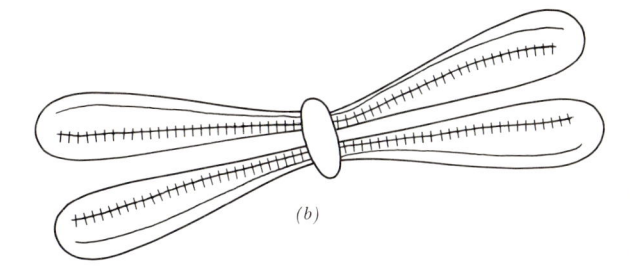

(b) At metaphase, the two sister chromatids are observed to be equally radioactive.

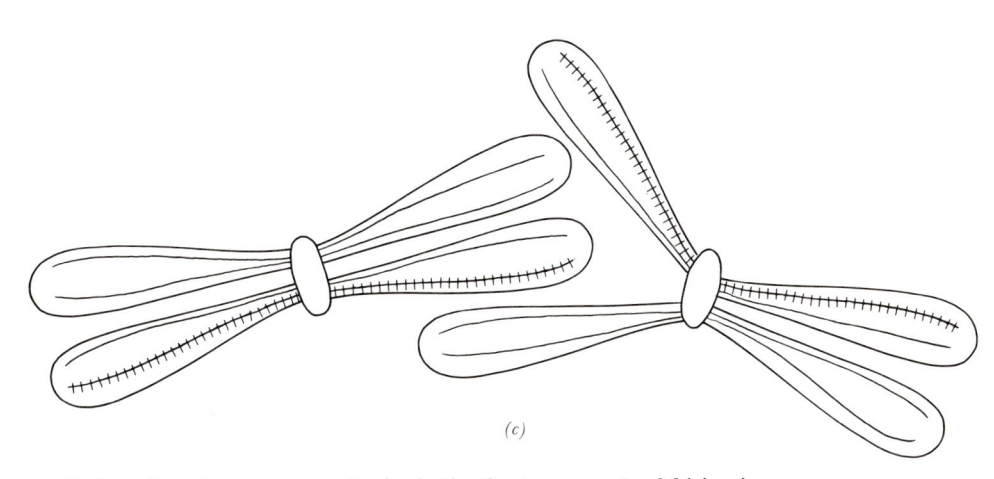

(c) If the cell undergoes a second mitosis (in the presence of colchicine in a nonradioactive medium), each of the daughter Chromosomes is observed to have one radioactive and one nonradioactive chromatid. The hot chromatid is observed to contain the same amount of label as did each of the chromatids in the previous metaphase.

duplication is semiconservative. Taylor's experiment is summarized in Fig. 5.9.

The actual arrangement of DNA in a Chromosome is as yet unknown. However, with respect to the transmission of atoms to daughters, Chromosomes duplicate as if they were single DNA molecules. The possibility that a single DNA molecule runs the entire length of a Chromosome would seem to be the model to rule out before expending too much thought on alternative schemes. Indeed, as techniques for extracting "protein-free" DNA become more gentle, the lengths of the individual DNA molecules recovered become longer—and there is no indication of an upper limit less than the total length of DNA per Chromosome. Autoradiography has been the most useful method for determining the length of long DNA molecules. This same method applied to DNA from Chromosomes labeled for short periods of time has revealed aspects of the organizational pattern of Chromosomes with respect to DNA duplication.

Organization of DNA with Respect to Duplication

In *E. coli,* autoradiography has revealed a simple rule for chromosome duplication; duplication begins at a point and progresses around the circular chromosome (Fig. 5.10). (In Chap. 8 we will tackle the question of how the *circular* E. coli chromosome is transferred as a *linear* molecule). In the isotope-transfer experiment of Meselson and myself (Chap. 3), the DNA examined in the centrifuge was (inadvertantly) fragmented upon extraction from the bacteria to pieces less than 1 percent the length of the whole chromosome. One generation after transfer of the culture to light-isotope medium *all* of the DNA fragments were half-heavy (Fig. 3.1); each short section of the chromosome had duplicated once and no section had duplicated twice. Grant the assumption that the transfer from N^{15} to N^{14} did not perturb the duplication of the chromosomes, and then convince yourself that this result, coupled with the autoradiograph, implies that duplication of the *E. coli* chromosome begins at the same point (and goes in the same direction!) in each generation.

A different sort of experiment supports the notion of sequential replication of the *B. subtilis* chromosome. In a nonsynchronized actively duplicating population of bacteria, twice as many individuals

FIG. 5.10. *An autoradiograph of a tritium-labeled chromosome of* E. coli. *This chromosome, which was caught in the act of duplicating, was extracted from a cell grown for two to three generations in medium containing tritiated thymidine. The regions of the chromosome labeled on one chain can be distinguished from those labeled on both by the density of ion tracks. The labeling pattern is diagrammed in the inset, solid lines standing for labeled and broken for unlabeled chains. This photograph was kindly supplied by* John Cairns.

100 μ

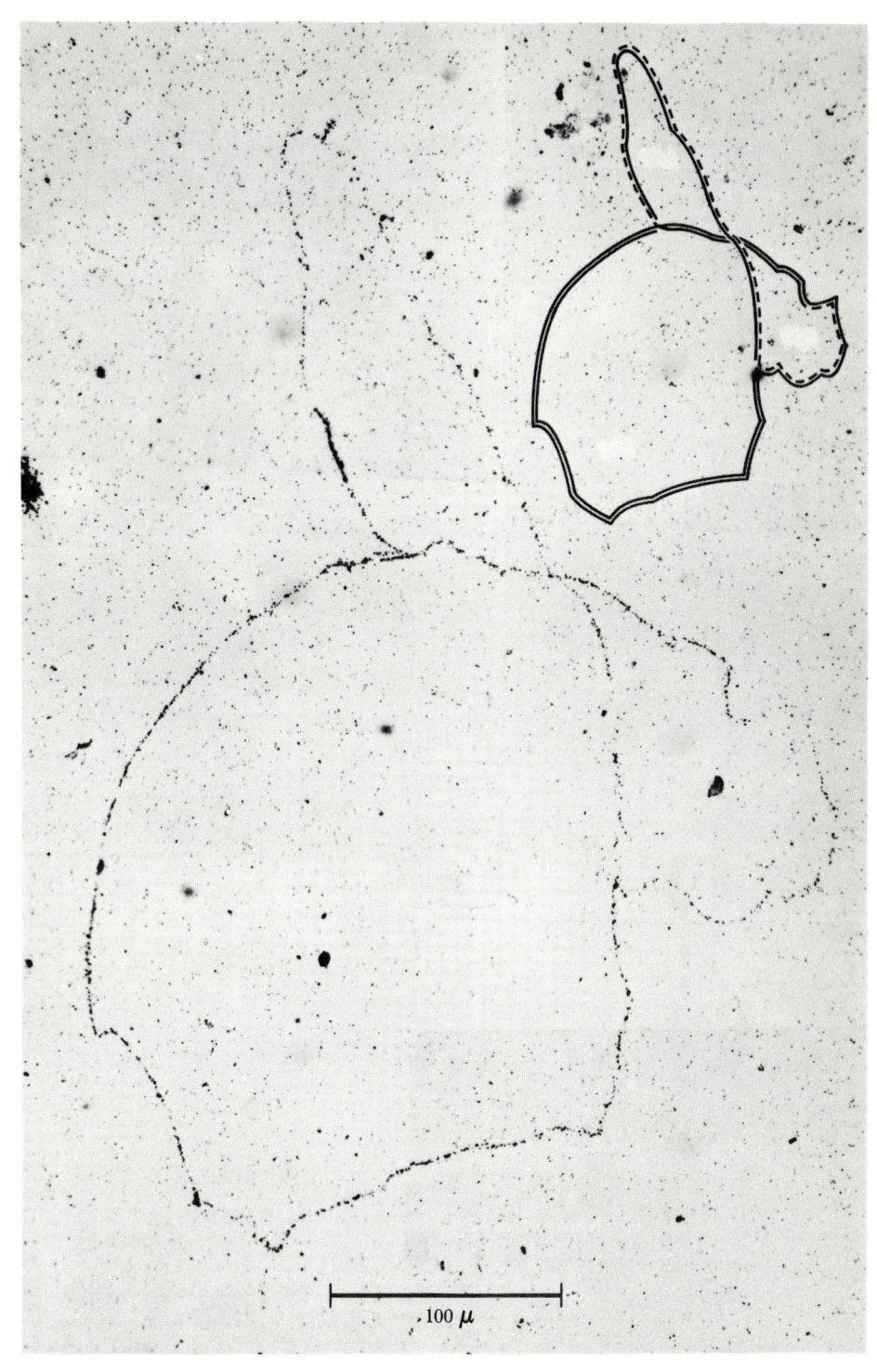

present at any time will have just undergone cell division as will be just about to undergo division. The time required to duplicate a bacterial chromosome under good growth conditions is known to be essentially equal to the generation time of the cells. Thus, the number of chromosomes which have just completed duplication, but have not yet embarked on the next cycle of duplication, is close to half the number which have just initiated duplication. Therefore, a stretch of DNA near the point of origin of duplication must be present in a population of exponentially growing bacteria in twice as many copies as a stretch near the terminus of duplication. Stretches in intermediate positions should be intermediate in numbers in the culture. Hiroshi Yoshikawa and Noboru Sueoka determined the relative number of copies of each of a number of stretches of DNA in *B. subtilis* by assaying the transformation capability for each of a number of hereditary characters of DNA extracted from exponentially growing cultures. The capabilities ranged over just a factor of two (when measurements were properly normalized). Thus an *order* could be assigned to the bits of DNA determining the characters transformed; bits near the origin are those which are present in the greater number of copies. The meaningfulness of the order so established was verified by the finding that characters with nearly equal transformation capability often acted as if they were on the same bit of DNA; a cell transformed for one of the characters was highly likely to be transformed for the other (see Problem 8.3).

In higher organisms autoradiographs have revealed more complex patterns of DNA duplication. The autoradiography has been carried out at two levels of analysis.

(1) Long DNA molecules isolated and radiographed after short periods of labeling have been found to bear labels in many places. Thus *each* molecule appears to have numerous points at which duplication is concurrently proceeding. The results of some labeling experiments are interpreted as meaning that duplication proceeds in both directions from each of a number of points of origin, like this:

(*o*'s mark origins and arrows mark direction of synthesis. New DNA chains are broken lines). Swivels of some kind would seem to be required for such a pattern of duplication. Perhaps single-chain interruptions like those identified in some viruses provide the necessary points of free rotation. Also plausible is the notion that the enzymatic replicating machinery itself introduces and repairs single-chain breaks as it moves along the Chromosome.

(2) In other experiments the distribution of radioactive thymidine label in intact metaphase Chromosomes is observed following short periods of label administered during the time of DNA synthesis in the

preceding interphase. These studies reveal that different Chromosomes, as well as different regions of individual Chromosomes, have characteristic times during interphase at which their DNA is synthesized. Since we are ignorant of the topography of the DNA within Chromosomes, the relation of these macroscopic labeling patterns to the microscopic ones described above is not known.

In viruses, the organization of DNA with respect to duplication is chaotic—both the situation and our state of understanding of it. For several viruses it nows seems likely that the duplicating DNA molecule is several (many?) times longer than the chromosome and that several (many?) points of duplication are simultaneously active. Genetic studies which relate to this question are described in Chap. 7.

Summary

In this chapter we have posed the problem of how DNA is organized within chromosomes. For viruses and bacteria, each chromosome seems to be a single nucleic acid molecule; the Chromosomes of higher organisms may contain only a few (perhaps one) long DNA molecules. Chromosomes function in an elongated state (intracellular phase of the virus life cycle, interphase in the mitotic cycle of higher organisms), and are packaged and distributed in a condensed state (mature virus particles, anaphase–telophase in the mitotic cycle). The topography of DNA in condensed chromosomes is not known.

References

De Robertis, E. D. P., W. W. Nowinski, and F. A. Saez, *General Cytology*, 3rd ed. Philadelphia: W. B. Saunders Co., 1960. A delightful classic in the field of cellular structure.

Jacob, François, and E. L. Wollman, *Sexuality and the Genetics of Bacteria*. New York: Academic Press, Inc., 1961. Chapters IX and XII are pertinent here, but all the chapters should be read eventually.

Taylor, J. Herbert, "The Time and Mode of Duplication of the Chromosomes," *Am. Naturalist, 91* (1957), 209–21. A clear, economically presented review of chromosome duplication with appropriate emphasis on the remarkable experiments of its author.

Thomas, C. A., Jr., "The Rule of the Ring," *J. Cell. Physiol., 70*, Supp. 1 (1967), 13–33. Nice experiments and excellent pictures reveal the variety of chromosome structure among viruses.

Problems

5.1. (a) If the chromosome of *E. coli* duplicates at the same rate per nucleotide as does the DNA of phage T2 (see Problem 3.3b), how much time is required for duplication? Compare your answer with the generation time of *E. coli* (20 min in nutrient broth at 37°C).

(b) How much time would be required to duplicate the DNA of *Drosophila* if we suppose that each of the *Drosophila* Chromosomes duplicates from one end to the other and that each of the Chromosomes duplicates during the same period?

5.2. About 90 min elapse between the onset of conjugation in *E. coli* and the entry into the female cell of the last genetic markers to be transferred.
(a) What is the length of chromosome (in microns) transferred per min?
(b) How long is an *E. coli* cell compared to the length of chromosome which it transfers in 1 min during conjugation?
(c) What provides the driving force for the transfer? (Make a guess, and then outline an experiment which might show your guess to be wrong.)

5.3. Chromosomes in root-tip cells of *Bellevalia* were labeled with H^3-thymidine. The roots were exposed to radioactive thymidine for a period sufficiently brief that no cells underwent more than one cycle of DNA duplication in the presence of label. The roots were then immersed in colchicine so that the number of postlabeling mitosis for each cell could be determined by chromosome counting.

Among cells which have undergone a postlabeling mitosis, an occasional chromosome is found that looks like this:

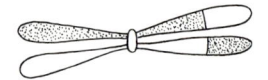

(The shaded portions indicate the presence of H^3.) These chromosomes appear to have exchanged homologous parts of sister chromatids. Sometimes one chromosome like the one above is found in a cell (call it a single exchange); sometimes *both* members of a pair of daughter chromosomes show the effects of exchange at the same level (call it a twin). A twin looks like this:

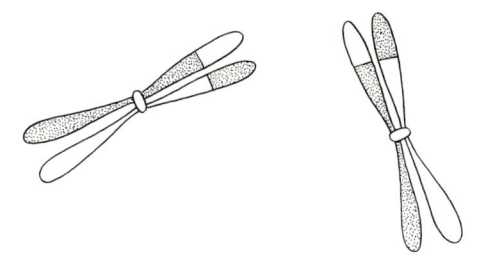

Let's focus on one chromosome that is permitted to duplicate in the presence of tritiated thymidine.
(a) At the first metaphase following labeling, how many of the two sister chromatids will be labeled with H^3?

(b) The two sister chromatids break at the same level and exchange parts shortly after labeling. (1) How many of the two sister chromatids will appear nonuniformly labeled at the first postlabeling metaphase? (Assume that the two subunits in each chromatid break at the same point and that these two subunits remain attached to each other throughout.) (2) How many of the four daughter chromatids at the second postlabeling metaphase will appear nonuniformly labeled if at the time of the exchange new (H^3) subunits always rejoin with old (H^1) subunits? (3) How many of the four daughter chromatids will appear nonuniformly labeled if at the time of the exchange new (H^3) subunits always rejoin with new (H^3) subunits? (4) How many of the four daughter chromatids will appear nonuniformly labeled if at the time of the exchange one of the exchanged pieces rejoins new to old and old to new while the other rejoins new to new and old to old?

(c) From chromatid exchanges occurring prior to the first metaphase what will be the ratio of singles to twins at second metaphase if rejoining of the exchanged parts is at random with respect to the age of the subunits?

(d) If a chromatid exchange occurs between the first and second metaphase, will it produce a single or a twin?

(e) What will be the ratio of events (detectable or not) occurring just prior to second metaphase to those occurring prior to first metaphase?

(f) At second metaphase: (1) What will be the ratio of singles to twins if reunion is always old to old and new to new? (2) What will be the ratio of singles to twins if reunion of exchanged parts is at random with respect to age of subunits? (3) What will be the ratio of singles to twins if reunion is always old to new and new to old? Taylor found a ratio of singles to doubles at second metaphase of about 2:1. The ratio was clearly less than 10:1. This result suggests that the rules governing reunion of chromatid subunits following exchange of a length of chromatid is such that old subunits always reunite with the new subunits of the sister chromatids. This rule suggests that the two subunits of a chromatid are not identical but are instead complementary. This evidence furthers the view that the subunits of a chromatid are the single chains of one DNA molecule or a linear array of DNA molecules.

Six

Recombination in Higher Organisms: I

In Chap. 2 we saw that a culture derived from a bacterial cell transformed from property X (say penicillin sensitivity) to property Y (say penicillin resistance) does not contain DNA that can transform Y cells to X. A bacterial cell can carry the DNA responsible for property X *or* property Y but generally not both. These two states of the bacterial chromosome are mutually exclusive alternatives; they are allelic to each other.

We have seen in Chap. 5 that the parts of a chromosome that affect each of a number of properties of a creature are located at distinctive places on a chromosome. When two creatures differ by a single mutation we shall call the distinctive bit of DNA in either of them a marker. The mutant marker is an allele of the original bit of DNA. Members of a set of alleles occur as mutually exclusive alternative markers at a distinctive locus on a chromosome.

When penicillin-sensitive, lactose-negative bacteria are exposed to DNA from penicillin-resistant, lactose-positive cells, the cells are transformed to drug resistance or to ability to ferment lactose but only occasionally to both states. If a drug-resistant, nonfermenting transformant is cultured, the DNA extracted from it can transform other cells to resistance (but not to sensitivity) and to the inability to ferment lactose (but not to the ability to ferment). A symmetric statement can be made regarding the drug-sensitive, lactose-positive transformants. Such transformed cells each have a chromosome not exactly like that of either the recipient or the donor strain but derivable by recombination of the genic material of the two strains.

We have abstracted the results of experiments on bacteria into a set of interrelated definitions in terms of the structure of the genic material set forth in Chaps. 1 to 5. In the discussion to follow there will not often be the opportunity to remind you of these definitions—in fact, the words will frequently be used without reference to either DNA or chromosomes. The formalisms of recombination analysis are self-contained; by completely independent logic and techniques they lead to a picture of a chromosome delightfully harmonious with the concept we have thus far developed.

Meiosis

Recombination requires that the genic material (or parts of it) from two or more different individuals be brought into intimate association. In higher organisms this is achieved by the processes of nuclear fusion and meiosis. Among the familiar organisms (cats, firs, and *Drosophila*) nuclear fusion occurs at the time of fertilization. Each of the gametes contains a haploid number of Chromosomes; the product of fertilization (the zygote) has twice as many, a diploid number, of Chromosomes. Many acts of mitosis follow fertilization; the resulting multicellular diploid individual (the cat, tree, or fly) becomes sexually mature, and some of its cells undergo meiosis, giving rise again to the haploid stage of the life cycle.

The familiar creatures have a prolonged, multicellular diplophase in contrast to many organisms that are more simple than these. For instance, in the green alga *Chlamydomonas*, meiosis follows nuclear fusion with no intervening acts of mitosis. In such simple creatures the primary role of meiosis is rendered obvious—it is the mechanism which provides for recombination. (We shall return to creatures with a prolonged diplophase in Chap. 11.)

The microscopically observable details of meiosis vary among the higher organisms. The Supplement that follows this chapter is a report by J. Kezer on meiosis as it occurs in salamanders. The photomicrographs in the Supplement are keyed to the diagrams of the corresponding generalized description of the stages of meiosis shown in Fig. 6.1.

As we did for the cases of mitosis, we can describe meiosis in terms of a sequence of changes in the appearance of the Chromosomes. Thus, we define the onset of meiosis as the time at which the Chromosomes have condensed sufficiently to be visible (with a light microscope). This stage, the leptotene stage of meiotic prophase (Fig. 6.1a), has been

FIG. 6.1. *A diagrammatic summary of the behavior of Chromosomes during meiosis.*

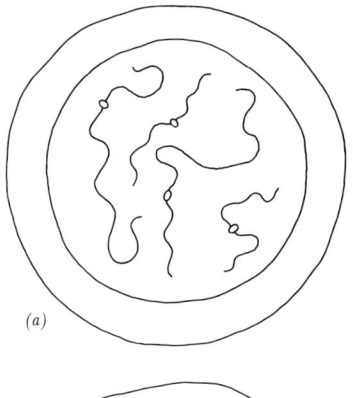

(a) **The Chromosomes first become visible to ordinary microscopic examination in the leptotene stage of meiotic prophase. Each Chromosome typically appears to be composed of but a single chromatid, though DNA duplication has occurred prior to this stage.**

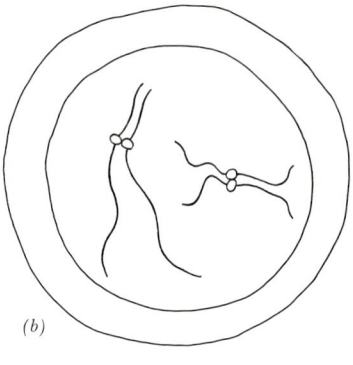

(b) **Homologous Chromosomes undergo pairing (synapsis) in the zygotene stage of meiotic prophase, producing Chromosomal structures called bivalents.**

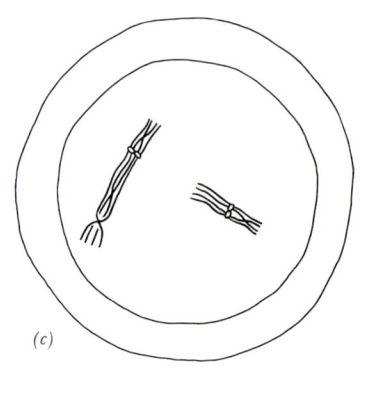

(c) **In many organisms (but not all of them, for example, the salamanders) each Chromosome is visibly composed of two chromatids at the pachytene stage of meiotic prophase, so that each bivalent can now be seen to consist of four strands of genic material.**

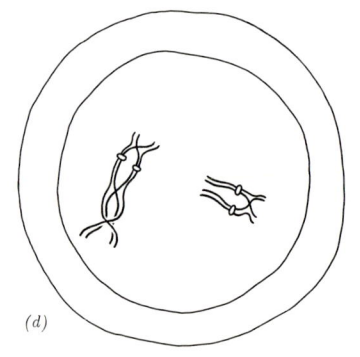

(d) **The chiasmata, which look as if they arose by exchange of homologous parts between synapsed chromatids, are clearly visible in the diplotene stage of meiotic prophase.**

(e) *Following the orientation of the four-strand bivalents (synapsed pairs of homologous, duplicated Chromosomes) on the equatorial plane of the spindle apparatus in metaphase I, the homologous centromeres move toward opposite poles of the spindle in anaphase I, separating the four strands of genic material two-from-two and thus producing two sets of half-bivalents.*

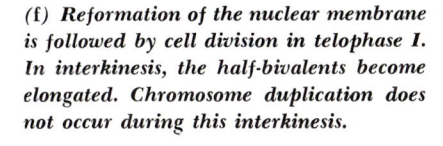

(f) *Reformation of the nuclear membrane is followed by cell division in telophase I. In interkinesis, the half-bivalents become elongated. Chromosome duplication does not occur during this interkinesis.*

(g) *The second meiotic division is initiated by shortening and thickening of the half-bivalents in prophase II followed by their migration to the equatorial plane of the spindle aparatus in metaphase II. The two chromatids making up each half-bivalent are separated one-from-one in anaphase II by the movement of sister centromeres toward opposite poles of the spindle.*

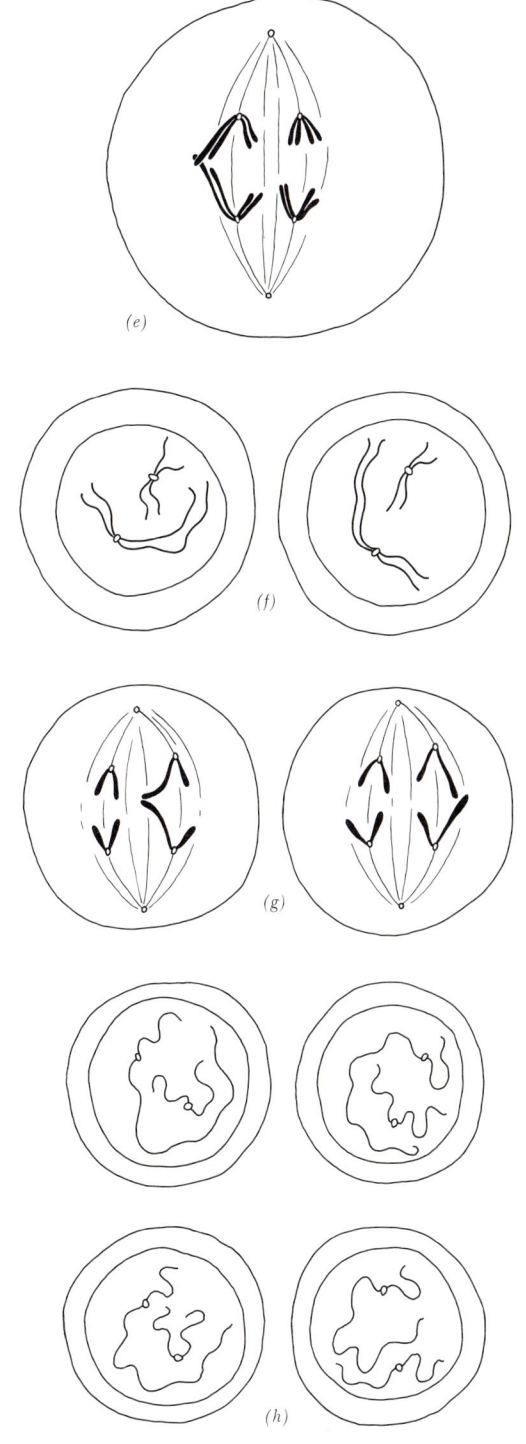

(h) *A second telophase results in the formation of four haploid cells, the final products of meiosis.*

97

preceded by a duplication of the DNA in each of the Chromosomes.

The Chromosomes look different from the way they look in the early prophase of mitosis; in fact, in many creatures, each Chromosome *appears* to be composed of but one chromatid; that is, even though the DNA has duplicated (as determined chemically), there is often no visible sign that the Chromosomes have duplicated. Homologous Chromosomes then enter into intimate pairing in the zygotene stage (Fig. 6.1b). The pairing (synapsis) begins often at one position on each pair of homologous Chromosomes and proceeds "zipperwise" to complete the synapsis. A synapsed pair of homologues is called a bivalent. The Chromosomes shorten and thicken, and each Chromosome divides longitudinally into two chromatids at the pachytene stage of meiotic prophase (Fig. 6.1c). In diplotene (Fig. 6.1d) the homologous Chromosomes appear to repel each other, but are typically held together by one or more chiasmata (the singular form is *chiasma*). These chiasmata *give the appearance* of having resulted from breakage, exchange, and reunion between a pair of chromatids, the two members of which are from different Chromosomes in the synapsed Chromosome pair.

The Chromosomes shorten and thicken and chiasmata often appear to move to the ends of the Chromosomes in the meiotic prophase stage called diakinesis. The nuclear membrane disappears and the pairs of Chromosomes held together by terminalized chiasmata assume positions on the equatorial plane of the spindle apparatus at metaphase I. Homologous centromeres with attached chromatids move to opposite poles of the spindle in anaphase I (Fig. 6.1e). The nuclear membranes form and the cell divides in telophase I. Each of the daughter cells then enters a period of interkinesis (an "interphase" without DNA synthesis; Fig. 6.1f) during which the Chromosomes become more or less elongated. This period is followed by shortening and thickening of the Chromosomes in prophase II. The Chromosomes, each composed of two chromatids, move to their respective equatorial planes in metaphase II. Centromeres divide and the two chromatids of each Chromosome move to opposite poles in anaphase II (Fig. 6.1g). The nuclear membranes reform and each cell divides. Four haploid cells (Fig. 6.1h) are the products of the meiosis.

Let us look now at the role of meiosis in the life cycle of the unicellular alga *Chlamydomonas*. Each *Chlamydomonas* cell contains one genome composed of several (about eight) nonhomologous Chromosomes, (the cells are haploid). The creature propogates by mitosis; one act of mitosis followed by cell cleavage gives two independent *Chlamydomonas* cells. Under certain conditions, related primarily to the nutrititional state of the cells, cells of opposite mating type unite in pairs. Fusion of the two cells is followed by fusion of the nuclei. There then ensues a meiosis which is probably similar to the generalized one described above. Following the completion of meiosis in *Chlamydomonas*, the spore case opens and the four cells, each free-swimming like the cells that originally fused, are liberated. The life cycle of *Chlamydomonas* is summarized in Fig. 6.2.

Segregation

In *Chlamydomonas* the two mating types are called mt^+ and mt^-. A given cell is of either one mating type or the other. When a *Chlamydomonas* cell divides, each of the daughter cells is of the same mating type as the parent. Mating type is "inherited." Among the four products of meiosis, two are almost invariably found to be of one mating type and two of the other. The two mating types, which are mutually exclusive, alternative properties, are manifested by the meiotic products *as if* they were determined by allelic markers, that is, as if they were determined by factors that segregate from each other at meiosis as do the homologous points of Chromosomes.

FIG. 6.2. *The life cycle of* **Chlamydomonas.** *Under appropriate conditions, haploid unicellular individuals of opposite mating types fuse in pairs. Nuclear fusion is followed by DNA duplication and meiosis wthin a spore case. The four haploid products, two of each mating type, are released upon germination of the zygospore. This figure, reproduced by permission of Paul Levine, appeared in* Cold Spring Harbor Symp. Quant. Biol., 23 *(1958), 102.*

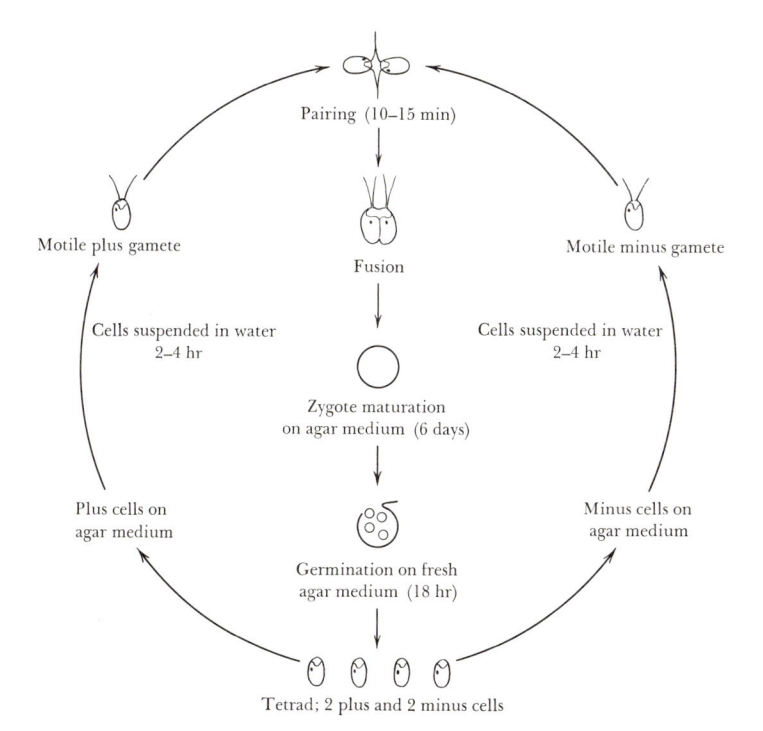

A number of mutant *Chlamydomonas* strains have been isolated through the use of techniques comparable to those described for bacteria in Chap. 1. Fusion and meiosis (mating) between mutant cells of one mating type and wild-type cells of the other mating type typically show the same phenomenon of 2 : 2 segregation. Two of the four products of each meiosis are wild type (like one of the parent cells) and the other two are mutant like the other parent.

Recombination

When mating occurs between *Chlamydomonas* cells differing by two or more hereditary properties (call this event a cross) the haploid products are frequently not all like one parent or the other with respect to all of the hereditary characteristics. We must conclude that the markers determining the various characteristics undergo recombination during meiosis.

Let us use symbols for the markers that differentiate the two parents in a cross. Let one parent carry a mutant marker at the a locus and a wild-type marker at the b locus, and designate it accordingly ab^+. Then let the second parent carry alleles to these markers; call it a^+b. The products of meiosis will generally be of four types:

$$\left.\begin{array}{l} ab^+ \\ a^+b \end{array}\right\} \text{parental types}$$

$$\left.\begin{array}{l} a^+b^+ \\ ab \end{array}\right\} \text{recombinant types}$$

It follows as a consequence of the 2 : 2 segregation of each of the pairs of alleles that among the four products (a tetrad) of any given act of meiosis

the number of ab^+ cells = the number of a^+b cells

and

the number of a^+b^+ cells = the number of ab cells

Thus, three kinds of tetrads can occur:

ab^+	ab^+	a^+b^+
ab^+	a^+b	a^+b^+
a^+b	a^+b^+	ab
a^+b	ab	ab

We can call them respectively

parental ditype tetrad (P)   tetratype tetrad (T)   recombinant ditype tetrad (R)

For crosses performed under standard conditions and involving a particular pair of loci the frequencies with which each of the three types of tetrads occurs is a reproducible characteristic property of that *pair of loci*. For most higher organisms, including *Chlamydomonas,* the most commonly encountered situation is that the frequency of P's equals the frequency of R's, and these two kinds of tetrads together make up a fraction of the total which is characteristic of that pair of markers. Since P's contain only products of parental type, R's contain only recombinants, and T's contain equal numbers of both types, these crosses give haploid products exactly half of which are recombinant. Markers behaving in this fashion are said to be unlinked. The equal frequency of P's and R's for unlinked markers can be simply understood if those markers are presumed to be situated at loci on *different* (nonhomologous) Chromosomes and if the orientation on the spindle at metaphase I of each bivalent were random. The *occurrence* of T's, however, is not accounted for by these correct, but inadequate, assumptions. They can be understood, on the other hand, if the chiasmata observed in meiotic prophase *do* reflect exchange between chromatids attached to separate homologous centromeres (nonsister chromatids). Convince yourself that a chiasma occurring in one bivalent between the marked locus and the centromere would be sufficient to account for the occurrence of tetratype tetrads.

Not all crosses give 50 percent recombinant types; occasionally, pairs of loci are found that give less. Such loci are said to be linked to each other. Vigorous collection of mutants and measurement of recombinant frequencies in many of the possible crosses involving two loci (two-factor crosses) reveal that groups of linked loci exist. Loci within any one group (a linkage group) are all demonstrably linked to each other (or to a common locus), whereas no linkage is demonstrable between any two loci in different linkage groups.

Crossing Over

As a rule, the data on recombinant frequencies between pairs of markers within a linkage group can be represented in the form of a one-dimensional linkage map. To construct a map, each marker is represented by a point in space. Two markers that show low frequencies of recombination with each other are located close together; markers giving large frequencies are put farther away from each other. When a number of markers have been located, the geneticist generally finds that he can draw one and only one nonbranching line from point to point (until all points lie on the line) such that the distance from one point to another *along the line* is invariably larger for markers that show large recombination values than for those showing smaller values.

The consistent success with which such efforts to construct one-dimensional maps have met suggests that the loci *do* exist on one-dimensional structures; the obvious candidates for the structures underlying the maps are the Chromosomes. (This conclusion is supported by an overwhelming body of evidence which you may examine in *Cytogenetics,* by Swanson, Merz, and Young, a volume in this series.) The events leading to recombination between linked markers are called crossovers (but see Chap. 9).

In crosses involving linked loci, tetratype tetrads are observed. It is clear therefore that crossing over cannot be accounted for solely by events occurring prior to the meiotic duplication of the genic material. Rather it *appears as if* crossing over at any one level occurs between only two of four strands. It seems likely that the strands are the four chromatids in each bivalent at prophase I and that the chiasmata are cytologically visible manifestations of crossing over.

The conclusion that crossing over occurs in the four-strand stage raised a set of complex problems: (1) Can any chromatid cross over with any other chromatid in a given set of four homologous chromatids? (2) Does the choice of chromatids engaging in crossing over at one level in the paired Chromosomes influence the choice at any other level? These questions have not been answered in a fully satisfactory fashion, but Problems 6.5 and 6.6 outline some of the evidence and the steps in the reasoning that leads to two tentative views: (1) Sister chromatids (the two chromatids resulting from the duplication of one Chromosome) do not cross over with each other. (2) The choice of chromatids engaging in crossing over at any one level has no influence on the choice at any other level; there is no chromatid interference.

The Mapping Function

The existence of one-dimensional linkage maps suggests that the frequency of recombinants for a given pair of loci is positively correlated with their physical distance apart on the Chromosome. This relation would follow if crossing over could occur at many different places along the paired Chromosomes. The quantitative relationships between recombination and crossing over can be examined from a study of measurements of the recombinant frequencies between pairs of loci for which the map sequences have been determined.

Granted the assumption of no sister-strand crossing over, all the tetrads arising from bivalents in which one crossover occurs will be tetratype. For bivalents with two crossovers in the marked region there are three ways in which the two crossovers may occur: (1) the two crossovers may involve the same two chromatids (in which case they are said to be regressive); (2) the two crossovers may involve one common chromatid only (progressive); (3) the two crossovers may involve completely different pairs of chromatids (digressive). These cases are diagrammed in Fig. 6.3. The members of a regressive pair of exchanges cancel each other and thereby produce a parental ditype tetrad. A progressive pair produces tetratype tetrads. A digressive pair

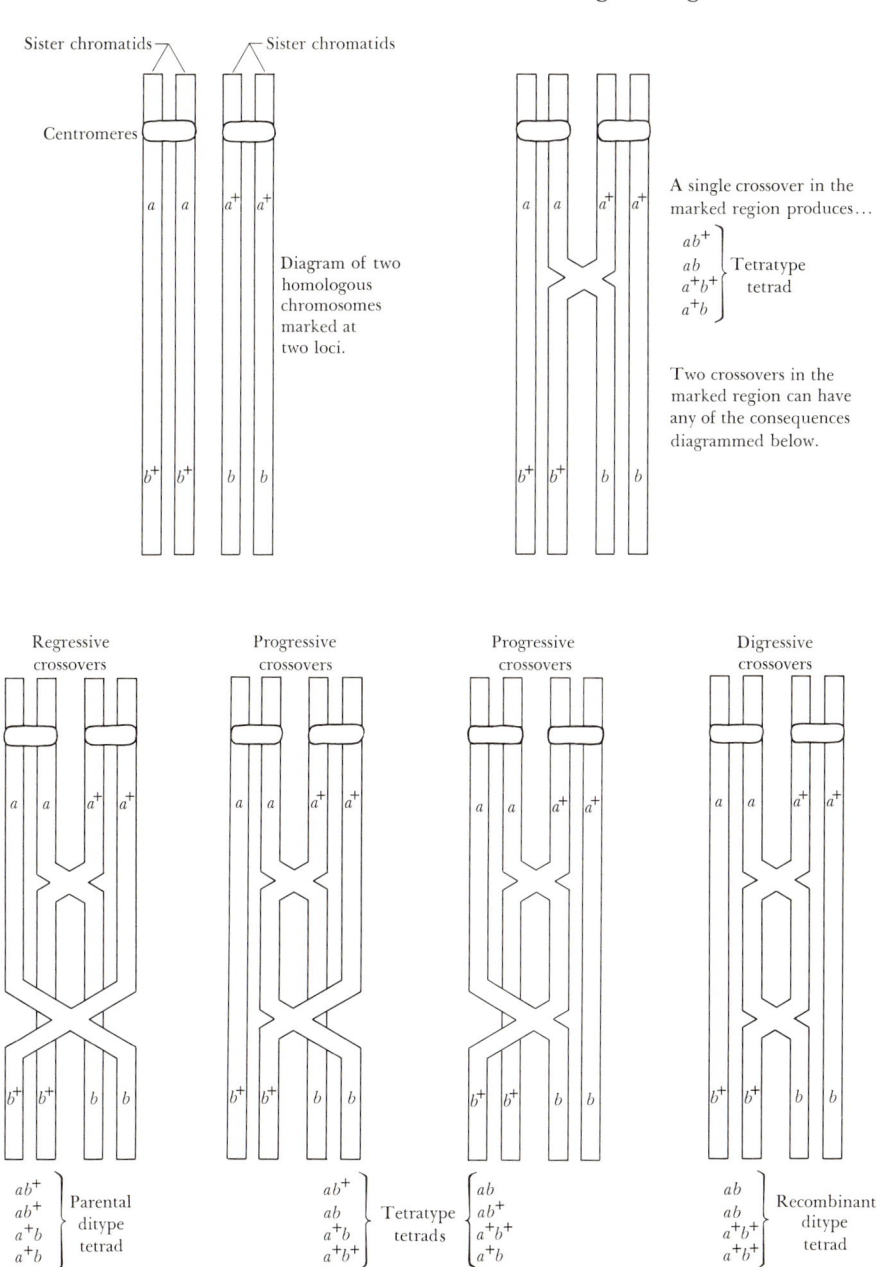

FIG. 6.3. *The manner in which two crossovers can dispose themselves in a bivalent. In the presumed absence of sister-strand crossing over, four distinguishable arrangements of two crossovers are possible. The four arrangements and their recombinational consquences are diagrammed here.*

of exchanges produces recombinant ditype tetrads. Since we have supposed an absence of sister-strand crossing over and of chromatid interference, these types will arise in the ratio

$\frac{1}{4}$P : $\frac{1}{2}$T : $\frac{1}{4}$R

For three exchanges the ratio will be

$\frac{1}{8}$P : $\frac{3}{4}$T : $\frac{1}{8}$R

(Convince yourself!)

In general, whatever the number of exchanges (greater than 0), P equals R. Since P's contain only products of parental type, R's contain only recombinants, and T's contain equal numbers of both types, it follows that the frequency of recombinants among tetrads with one or more crossovers between two marked loci will be $\frac{1}{2}$.

We are now able to write a null hypothesis that relates the average number of crossovers between two marked loci with the frequency of recombinants. Let us *assume* that among the bivalents the numbers of crossovers are distributed according to the Poisson expression (see appendix on Poisson distribution). This assumption is equivalent to the assumption that the presence of one crossover neither increases nor decreases the likelihood of a second crossover; crossovers are assumed to be independent events. (We shall see later that this is rarely true.) Then, if the average number of crossovers between two marked loci is called x, the fraction of bivalents with no crossovers between the marked loci is $x^n e^{-x}/n!$ evaluated at $n = 0$, or, simply, e^{-x}. The fraction of bivalents with one or more crossovers between the marked loci is, therefore, $1 - e^{-x}$, and the frequency of recombinants is

$p = \frac{1}{2}(1 - e^{-x})$ (Eq. 6.1)

Whenever x is small ($\ll 1$), $(1 - e^{-x}) = x$ (approximately), and

$p = x/2$ (approximately, when x is small) (Eq. 6.2)

We wish now to define the "distance" d between two loci such that d is proportional to x and for closely linked loci is equal to p. Assuming Eqs. 6.1 and 6.2 we can write $d = x/2$ and

$p = \frac{1}{2}(1 - e^{-2d})$ (Eq. 6.3)

If we now define a *map* distance as $D = 100d$, we can write

$p = \frac{1}{2}(1 - e^{-2D/100})$ (Eq. 6.4)

Equations 6.1, 6.2, and 6.4 are plotted in Fig. 6.4.

An inspection of Fig. 6.4 reveals several properties of an "ideal" map.

(1) Throughout the range where the curve is approximately a straight line with a slope of one, recombinant frequencies equal "distance" and are, therefore, additive. For three closely linked loci, 1, 2, and 3 linked in that order, the recombinant frequency from a cross in-

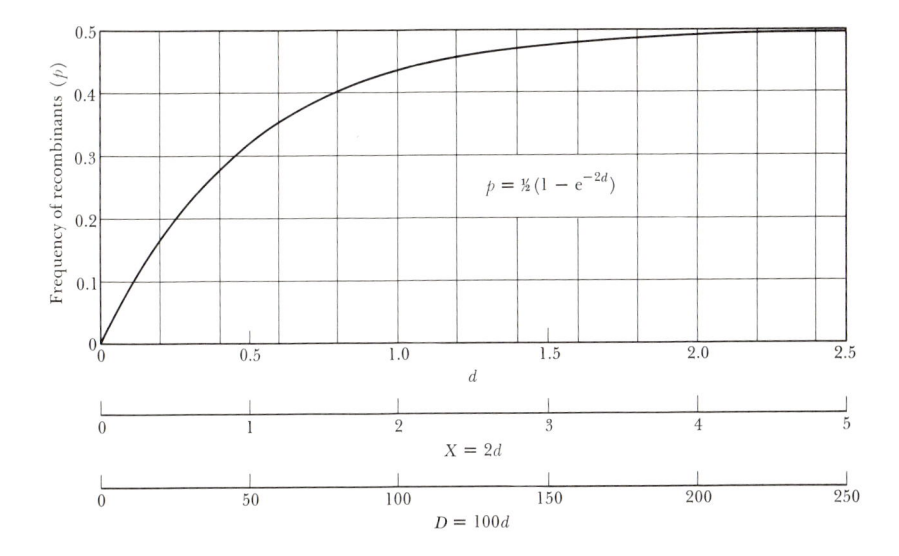

FIG. 6.4. *The mapping function for recombination in higher organisms in the absence of interference. The recombinant frequency (p) increases linearly with map distance (D) at short distances. At long distances the recombinant frequency asymptotically approaches 0.5 and widely separated markers then appear to be "unlinked." The three metrics on the abscissa refer to Eqs. 6.3, 6.1, and 6.4 respectively.*

volving 1 and 3 equals the sum of the recombinant frequencies from the two crosses involving 1 and 2 and 2 and 3 respectively.

(2) In the range where the curve has appreciable curvature, recombinant frequencies are not additive. When the map distances between three linked loci 1, 2, and 3 are rather large, the frequency of recombinants from a cross involving 1 and 3 is less than the sum of the recombinant frequencies from the two crosses involving 1 and 2 and 2 and 3 respectively.

(3) When map distances between markers are very large, the frequency of recombinants is independent of map distance and equals $\frac{1}{2}$.

Departure from the Ideal Mapping Function

With most organisms the recombinant frequencies depart to some degree from the "ideal" mapping function of Fig. 6.4. If our assumptions regarding the absence of sister-strand crossing over and chromatid interference are correct, these departures must represent inadequacies in our assumption of a Poisson distribution of crossovers among bivalents.

For any trio of linked markers, departures from ideality, if they exist, can be detected as follows. Recombinant frequencies for each of the crosses 1 × 2, 2 × 3, and 1 × 3 are measured experimentally. These recombinant frequencies are then converted to "ideal" map distances. This conversion can be made conveniently by reading from the graph in Fig. 6.4. The map distances are then tested for additivity; the map distance between loci 1 and 3 is compared with the sum of the map distances between 1 and 2 and 2 and 3 respectively. If recombinant frequencies *do* depend on map distance in an ideal fashion, the derived map distances will indeed be additive.

Additivity may fail, however, for either of two reasons: the "ideal" map distance between loci 1 and 3 may come out to be either greater than *or* less than the sum of the two map distances for loci 1 and 2 and 2 and 3 respectively. In most organisms, systematic deviations from additivity are commonly found. For rather large recombinant frequencies the calculated map distance for loci 1 and 3 is typically greater than the sum of the calculated map distances for the other two pairs of loci. (In Chap. 7 we shall see that additivity usually fails in the other direction among viruses.)

Three-Factor Crosses

A quantitative measure of the deviations from a Poisson distribution of crossovers among synapsed homologues can be conveniently expressed in terms of the frequencies of types arising from a cross involving three marked loci (a three-factor cross) all of which are linked to each other. Call the three loci 1, 2, and 3, and let them be marked by the alleles a and a^+, b and b^+, c and c^+ respectively. If the two parental types are abc and $a^+b^+c^+$, then the emerging types and their designations may be written as follows:

$$\left.\begin{array}{l} abc \\ a^+b^+c^+ \end{array}\right\} \text{parental}$$

recombinants for loci 1 and 2 = p_{12} $\left\{\begin{array}{l} \left\{\begin{array}{l} ab^+c^+ \\ a^+bc \end{array}\right\} \text{recombinants for loci 1 and 2 only} \\ \left\{\begin{array}{l} ab^+c \\ a^+bc^+ \end{array}\right\} \text{double recombinants} \\ \left\{\begin{array}{l} abc^+ \\ a^+b^+c \end{array}\right\} \text{recombinants for loci 2 and 3 only} \end{array}\right\}$ recombinants for loci 1 and 3

recombinants for loci 2 and 3 = p_{23}

If crossovers are Poisson distributed, then recombination for loci 1 and 2 and for loci 2 and 3 will be statistically uncorrelated events. Under this condition, the frequency of double recombinants would be

$p_{12} \times p_{23}$. We may define S, the coefficient of coincidence, by the equation

Actual frequency of double recombinants = $S\ p_{12}p_{23}$ (Eq. 6.5)

The frequency of recombinants for loci 1 and 3 may then be written in terms of p_{12}, p_{23}, and S as

$p_{13} = p_{12} + p_{23} - 2S\ p_{12}p_{23}$ (Eq. 6.6)

We see that in principle S can be estimated experimentally either from the results of three-factor crosses by application of Eq. 6.5 or from the pooled results of each of the three two-factor crosses by application of Eq. 6.6. Values of S less than 1 mean that there is a tendency for adjacent pairs of loci not to undergo simultaneous recombination. Such an experimental result would indicate that crossovers tend to interfere with each other. Values of S greater than 1 would mean that adjacent pairs of loci undergo simultaneous recombination more often than would be expected if crossovers were Poisson distributed among synapsed homologues.

In the discussion above it was noted that calculated "ideal" map distances for moderately linked markers fail to be additive because D_{13} as estimated from the ideal mapping function is greater than the sum of D_{12} plus D_{23}. This statement is equivalent to the statement that

$p_{13} > p_{12} + p_{23} - 2p_{12}p_{23}$

which implies $S < 1$.

The simplest interpretation of values of S less than unity is that the presence of a crossover between loci 1 and 2 discourages the occurrence of a crossover between loci 2 and 3, that is, there is "interference" between crossovers. Studies of patterns of interference have not shed much light on the mechanism of crossing over. Interference is generally attributed to an as yet ill-defined "mechanical property" of the Chromosomes. As long as geneticists performed crosses involving markers that were not very close together, the crossover model as we have described it above provided a consistent framework for linkage analysis. Complications arose, however, when analysis was extended to the study of recombination between very closely linked markers. We shall examine these complications in Chap. 9 and consider a model for recombination between linked markers which encompasses them.

Summary

In most higher organisms the genic material is organized into more than one Chromosome. Recombination of markers at different loci can, therefore, come about in either of two ways. Markers at loci on different Chromosomes assort independently as a consequence of the random orientation of bivalents at the metaphases of meiosis. Markers at different loci on the same Chromosome recombine by crossing over.

Crosses involving linked markers generate data, measured as frequencies of recombinants, which permit the construction of genetic linkage maps. These maps accurately reflect the order of the markers on the Chromosomes (see *Cytogenetics* by Swanson, Merz, and Young, in this series, for exceptions) and support the idea of the fundamental linearity of the Chromosomes.

References

Barratt, R. W., Dorothy Newmeyer, D. D. Perkins, and Laura Garnjobst, "Map Construction in *Neurospora crassa*," *Adv. Genet.*, 6 (1954), 1–93. From tetrad analysis and analysis of random meiotic products the authors develop mapping functions for *Neurospora*. The functions, which relate tetratype tetrad frequencies or recombinant frequencies with map distance, take into account the existence of interference. Mathematically inclined students might enjoy and profit from this paper.

Creighton, H. B., and B. McClintock, "A Correlation of Cytological and Genetical Crossing over in *Zea Mays*," *Proc. Natl. Acad. Sci. U.S.*, 17 (1931), 492–97. Reprinted in *Classic Papers in Genetics*, J. A. Peters, ed. (Englewood Cliffs, N.J.: Prentice-Hall, Inc., 1959), pp. 155–60.

Sturtevant, A. H., "The Linear Arrangement of Six Sex-linked Factors in *Drosophila*, as Shown by Their Mode of Association," *J. Exp. Zool.*, 14 (1913), 43–59. Reprinted in *Classic Papers in Genetics*, J. A. Peters, ed. (Englewood Cliffs, N.J.: Prentice-Hall, Inc., 1959), pp. 67–78.

Problems

6.1. Suppose we throw into a pot a large number of *Chlamydomonas* haploid cells of one genotype [for instance, able to synthesize nicotinamide (nic^+) and an equal number of cells unable to make nicotinamide (nic^-)] and of the opposite mating type. The two *nic* markers behave as alleles. Cell and nuclear fusion are allowed to occur. The resulting cells are diploid. We now perform the following experiments on this pot.

(a) We remove a single diploid cell, permit meiosis to occur, and examine the 4 haploid products. How many are nic^+?

(b) We remove 10 diploid cells, permit meiosis to occur, and examine all 40 emerging products. How many are nic^-?

(c) We permit the very large number of diploid cells in the pot to undergo meiosis and we let the haploid products swim about until the contents of the pot are randomized. We pick one cell at random. What is the probability that it is nic^+?

(d) Suppose a locus *mt* is on a different Chromosome from the *nic* locus. What fraction of the cells in (c) will be mt^+ nic^-? What fraction of the cells will be recombinant types?

In many higher organisms (for example, cats, firs, and flies) the four products of individual acts of meiosis are not recovered together. Instead, the haploid cells (gametes for the cats and flies, gametophytes for the firs) from many separate acts of meiosis are in one sense or another tossed into a common pot. The *definition* of an unlinked pair of markers in this case is that the frequency of recombinant types is not detectably less than 50 percent and that the two markers do not both show linkage to any other single marker or set of linked markers. By inference, we *would* find $P = R$ for these markers if we *could* analyze their tetrads as we can for *Chlamydomonas*. The genetics of cats and flies is further covered in Chap. 11.

6.2. (a) What four kinds of meiotic (haploid) products can be produced by a diploid cell that contains the alleles A and a and the alleles B and b?

(b) Assume that the diploid cell in Problem 6.2a arose from the union of two haploid cells of genotypes AB and ab respectively. (1) Of the 4 kinds of meiotic products, which are "parental" and which are "recombinant"? (2) In an experiment, suppose that the following numbers of meiotic products are counted:

| AB | 221 | aB | 64 |
| ab | 232 | Ab | 66 |

What is your estimate of the recombinant frequency from a cross involving the loci a and b? To how many "ideal" units does this correspond?

(c) Suppose the diploid cell in Problem 6.2b2 is also segregating the alleles C and c. (1) How many genotypes of meiotic products can be produced? Specify them. (2) Suppose that the frequency of recombinants for loci b and c is 20 percent and for a and c is 38 percent. What is the map sequence of the loci? Calculate the coefficient of coincidence for the regions ab and bc. In a three-factor cross involving loci a, b, and c, what frequency of double recombinants would you expect to find?

6.3. Occasionally, pieces of one Chromosome can become attached to the end of another Chromosome or even become inserted into another Chromosome. Such an event is a translocation. Suppose two loci a and b are known. A piece of another Chromosome has become inserted into the region between a and b. Crosses in which neither parent carries the translocation give 25 percent recombinants for the loci a and b. Crosses in which both parents carry the translocation give 38 percent recombinants for the same two loci. If you assume the applicability of the "ideal" mapping function, what is the map length of the translocated piece of Chromosome?

6.4. Following X irradiation, aberrant Chromosomes can be found in increased frequency. "Ring" Chromosomes are found in which the two ends of one Chromosome have fused to each other so that the Chromosome looks like this:

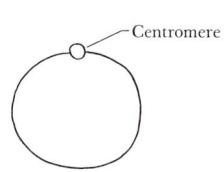

During meiosis in a cell carrying two such ring Chromosomes (and no normal homologue), the rings synapse with each other and undergo crossing over in the four-strand stage. At anaphase I, any bivalent in which one of the chromatids has engaged in an odd number of nonsister-strand exchanges gets "hung up" since the homologous centromeres are connected to each other. An odd number of sister-strand exchanges would be expected to show such "bridges" at anaphase II, but bridges at anaphase II do not seem to occur. What does this tell us about sister-strand crossing over?

6.5. (a) Assume that no crossing over between sister chromatids can occur. Among chromatid tetrads in which *one* crossover between two loci has occurred, what frequency of tetratype spore tetrads will result? (b) Assume that sister chromatid crossing over is just as likely to occur as any other kind. (1) Among tetrads in which *one* crossover has occurred between loci, what frequency of tetratype tetrads will result? (2) Among tetrads in which 2 (or 3, or 4) crossovers have occurred, what frequency of tetratype spore tetrads will result if there is no chromatid interference? if there is complete positive chromatid interference?[1]

(c) In fact, situations are found in which there are more than 66.7 percent tetratype tetrads. What does this tell us about sister-strand crossing over and chromatid interference?

6.6. In some fungi of the order Ascomycetes, meiosis proceeds in a long thin sac (ascus). The nuclei can seldom slip past each other so that the terminal member at either end of a tetrad and the cell next to it are sister cells of the same second meiotic division. This situation permits the recognition of recombination between marked loci and the centromere. Suppose a locus is marked by the alleles a and a^+. If recombination has not occurred the sequence of markers in the tetrad will be either aaa^+a^+ or a^+a^+aa.

(a) What will be the relative frequencies of these two types of tetrads?

If recombination has occurred, it *may* result in the appearance of tetrads with different marker sequences, for example, aa^+aa^+, a^+aaa^+, or aa^+a^+a. Such sequences indicate that the alleles a and a^+ have, because of crossing over, segregated in the second meiotic division. Assume the validity of the notion that there is no sister-strand crossing over and no chromatid interference.

(b) Suppose that in a particular bivalent *one* crossover has occurred between locus a and the centromere. What is the probability that the tetrad will manifest second division segregation?

(c) Suppose exactly two crossovers have occurred between the locus a and the centromere. What is the probability that a and a^+ will segregate in the second division? If exactly three crossovers occur? If a vast number of crossovers occur?

---

[1] Positive chromatid interference is the situation wherein an adjacent pair of crossovers tend to involve different chromatids. In the case of complete positive interference, two adjacent crossovers invariably involve no common chromatid.

Assume now that sister-strand crossing over does occur but again that there is no chromatid interference.

(d) What is the probability that the tetrad will manifest second division segregation if in a particular bivalent *one* crossover occurs between locus *a* and the centromere? If two or more crossovers occur? Under what conditions using an ascomycete might you rule out the combined hypothesis of sister-strand crossing over with no chromatid interference?

Supplement to Chapter 6

Meiosis in Salamander Spermatocytes
by James Kezer
Department of Biology, University of Oregon

The large size of salamander Chromosomes and the ease with which they can be prepared for observation make these amphibians ideal as a source of cells for the study of the meiotic process. The accompanying plates of photomicrographs show the events of meiosis as they occur in certain members of the Plethodontidae, a family of salamanders abundantly represented in the United States, Mexico, Central America, and northern South America. A single genus, *Hydromantes,* occurs in Europe, in southern France, northern Italy, and Sardinia.

To understand the events of meiosis that are shown in the photomicrographs, it is necessary to recall that all of the cells of a mature salamander are derived by mitosis from a fertilized egg, the zygote. In salamanders, as in other organisms, a fusion of sperm and egg nuclei takes place during fertilization, producing a zygote nucleus that contains two intimately associated sets of Chromosomes from different salamanders. As a result, the Chromosomes of the zygote exist in pairs: for each Chromosome brought in by the sperm nucleus, there will be a Chromosome similar in appearance contributed by the egg nucleus. In many of the plethodontid salamanders, but not in all of them, there are 28 Chromosomes in the zygote, 14 brought in by the sperm and 14 by the egg. One member of each Chromosome pair is said to be homologous to the other member of the pair: for salamanders in which $2n = 28$, the zygote nucleus will contain 14 pairs of homologous Chromosomes, and all of the thousands of somatic cells derived from the zygote by mitosis will contain replicas of these 14 homologous pairs.

Meiosis occurs in those cells that are the immediate ancestors of the gametes. In vertebrates and other multicellular animals, these are the spermatocytes of the testes and oocytes of the ovaries. Meiosis consists of two cellular divisions, designated very simply and appropriately as the first and second meiotic divisions.

Consider, for example, the events that occur in the cells that make up the germinal epithelium of a salamander testis. The cells of this epithelium, called spermatogonia, grow and divide by mitosis, producing other cells like themselves. At a particular time during the year, mitotic activity ceases, the spermatogonia increase in size and become spermatocytes, ready for the sequence of events that will transform them into mature gametes. The first of these events consists of the two meiotic divisions which (1) reduce the Chromosome number by one-half, (2) separate the homologous Chromosomes and segregate them in different nuclei, and (3) bring about a recombination of the genic material of homologous Chromosomes. We will now follow a spermatocyte through the two meiotic divisions, using the accompanying plates of photomicrographs, to see how these three events are accomplished.

Figure 6S.1 shows an unfixed and unstained salamander spermatocyte just entering the first meiotic division. It is in a tissue culture chamber, somewhat compressed beneath a strip of dialysis membrane, and has been photographed with phase contrast optics. The large oval nucleus contains an almost centrally located nucleolus. Although the Chromosome strands cannot be seen, the parts of the Chromosomes associated with the centromeres are visible as small dark granules scattered throughout the nucleus. The cytoplasm contains many large, dense granules and rodlike mitochondria that are not clearly in focus. The conspicuous spherical structure in the cytoplasm is a centrosome, surrounded by a layer of compact granules which can be shown by electron microscopy to be Golgi membranes. A pair of centrioles (not visible in the photomicrograph) is imbedded in the centrosome. During prophase of the first meiotic division, this centrosome divides. The two resulting centrosomes, with their centrioles, move to the poles of the meiotic apparatus from which the spindle and the asters radiate. This most extraordinary photomicrograph was obtained from a preparation made by Dr. Takeshi Seto and myself in connection with our study of meiosis in living, in vitro salamander spermatocytes.

The other photomicrographs were obtained from aceto-orcein squash preparations of salamander spermatocytes, made in connection with my research on the cytoevolution of the plethodontid salamanders. To obtain preparations such as those illustrated in the photomicrographs,

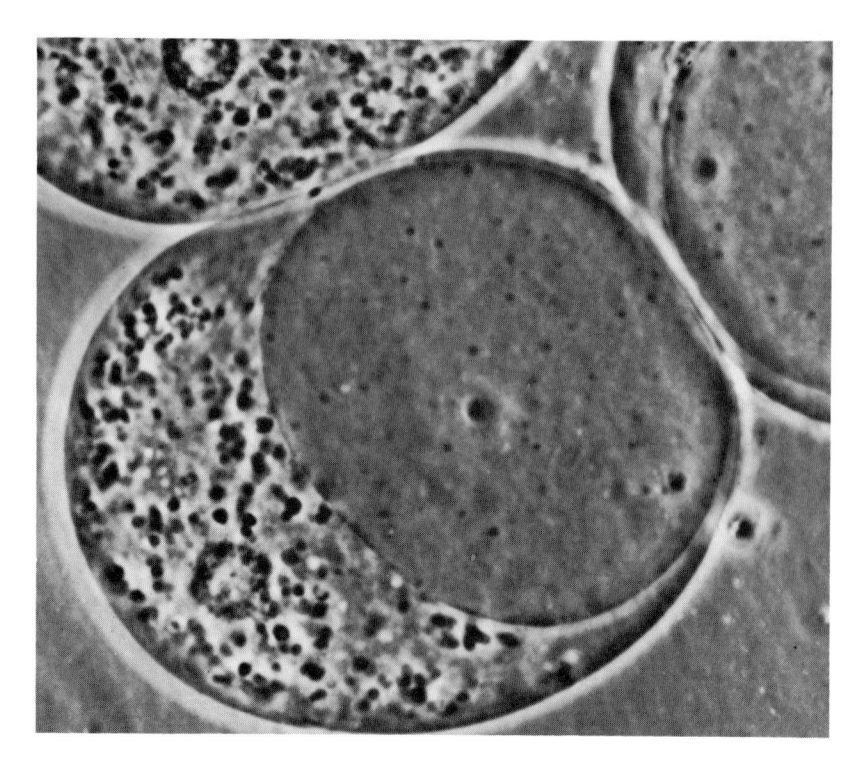

FIG. 6S.1. Batrachoseps wrighti: *a living spermatocyte, in vitro, photographed with phase contrast optics.*

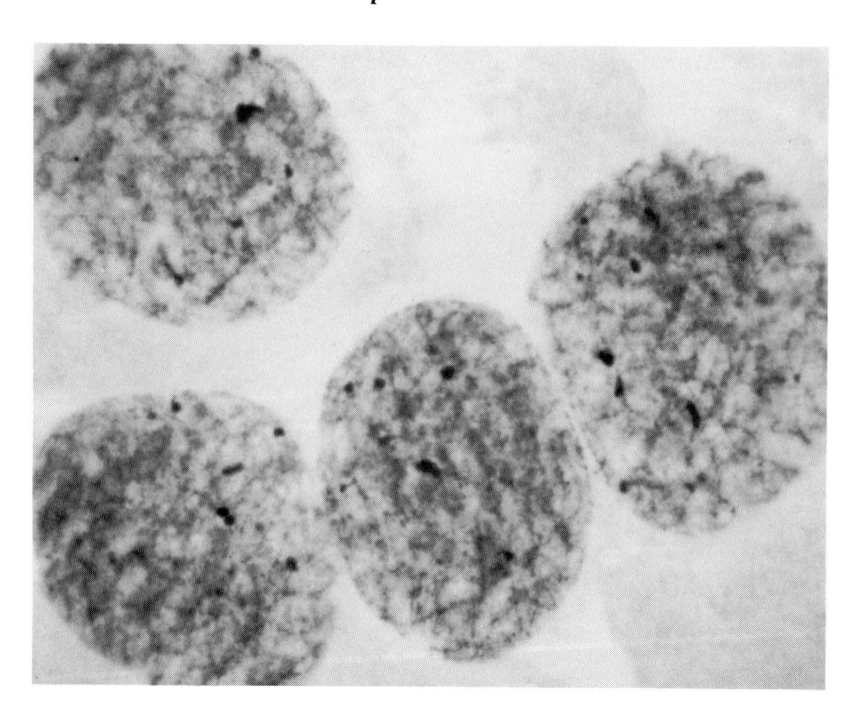

FIG. 6S.2. **Plethodon vehiculum.** *Leptotene.*

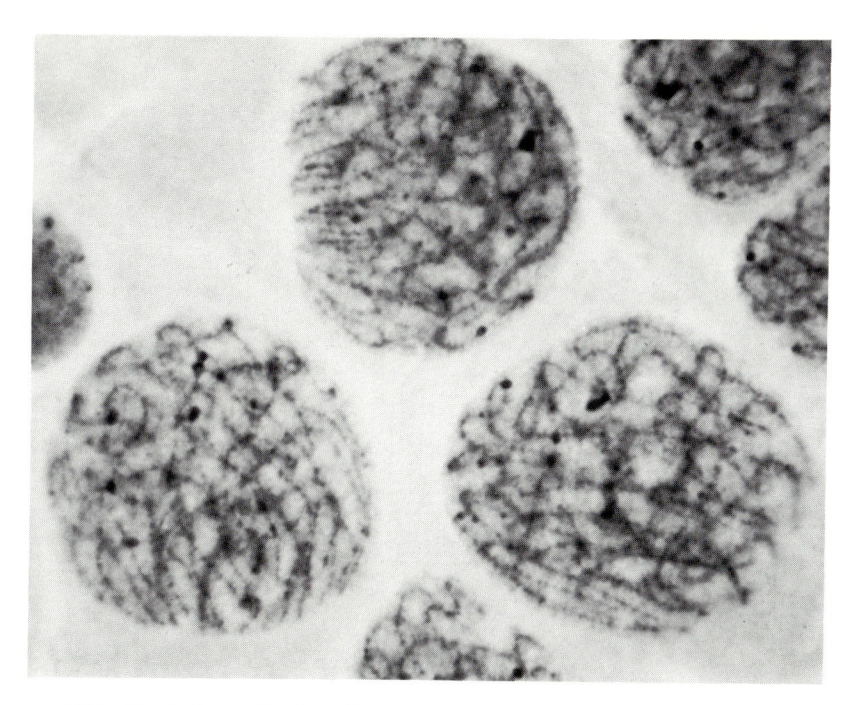

FIG. 6S.3. **Plethodon vehiculum.** *Zygotene.*

FIG. 6S.4. **Plethodon vehiculum.** *Pachytene.*

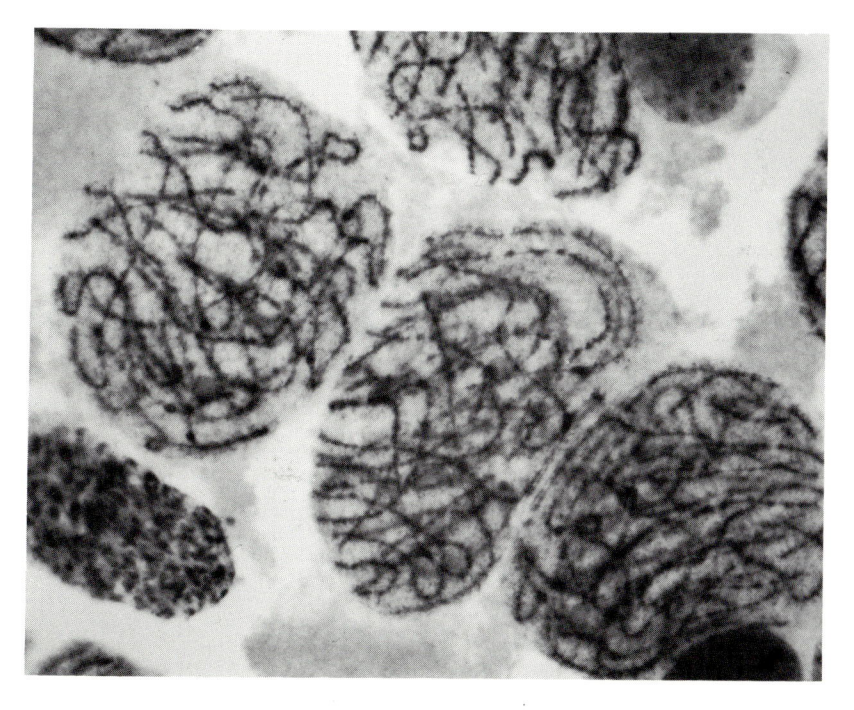

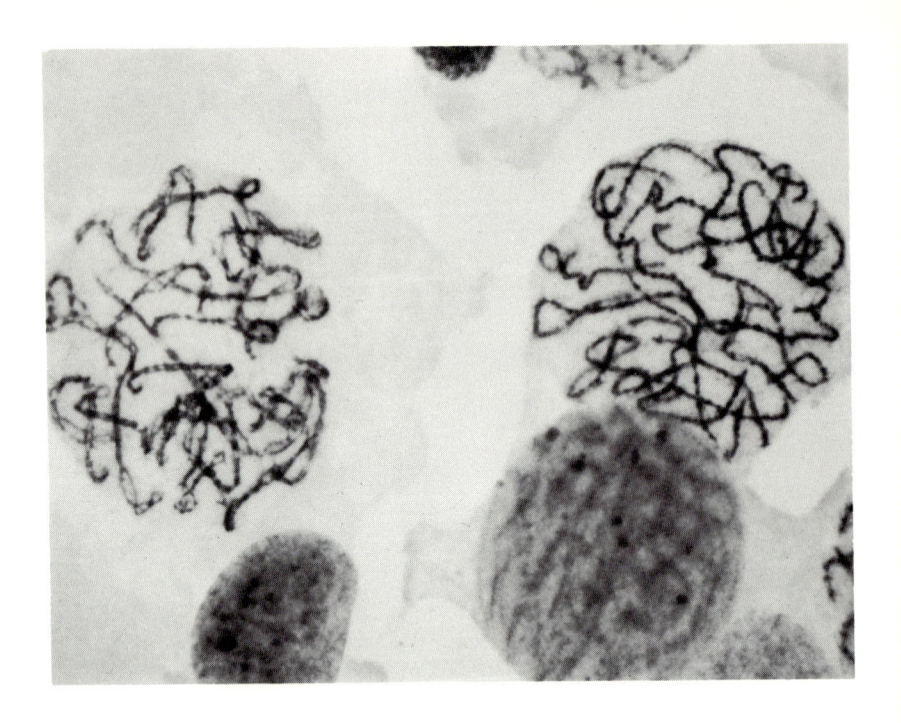

FIG. 6S.5. **Plethodon vehiculum:** *Early diplotene nucleus on the left; pachytene nucleus on the right.*

FIG. 6S.6. **Plethodon vehiculum:** *Early diplotene nucleus on the left; zygotene nucleus on the right.*

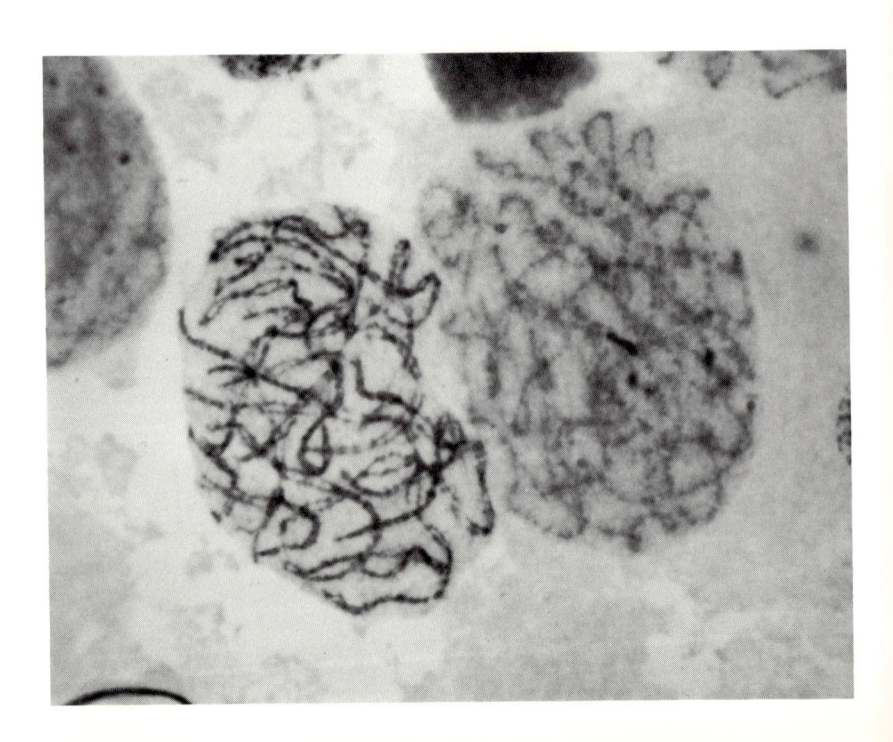

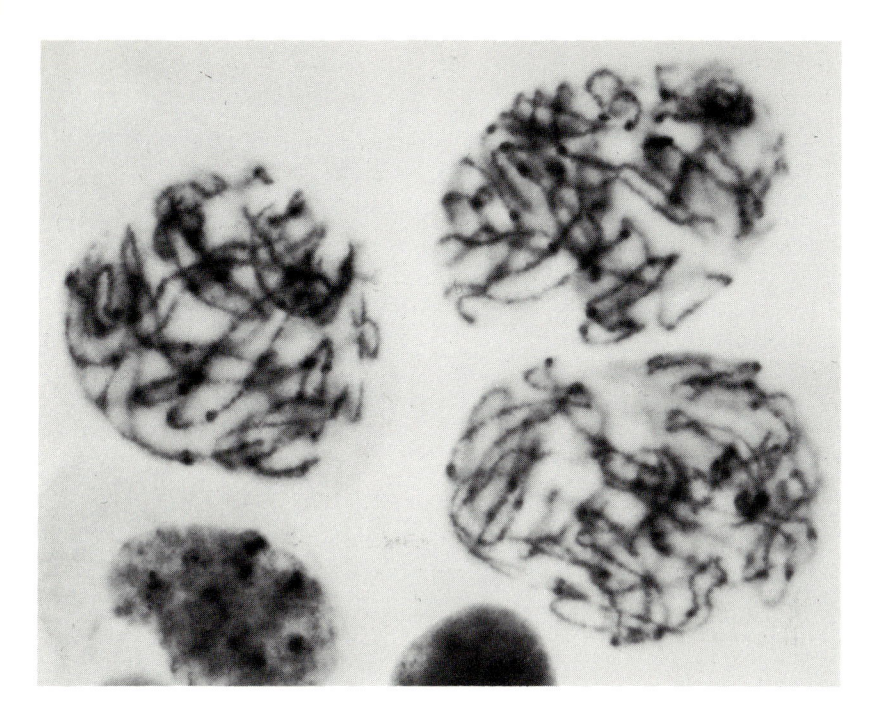

FIG. 6S.7. **Plethodon cinereus cinereus.** *Diplotene.*

FIG. 6S.8. **Plethodon cinereus cinereus.** *Diplotene.*

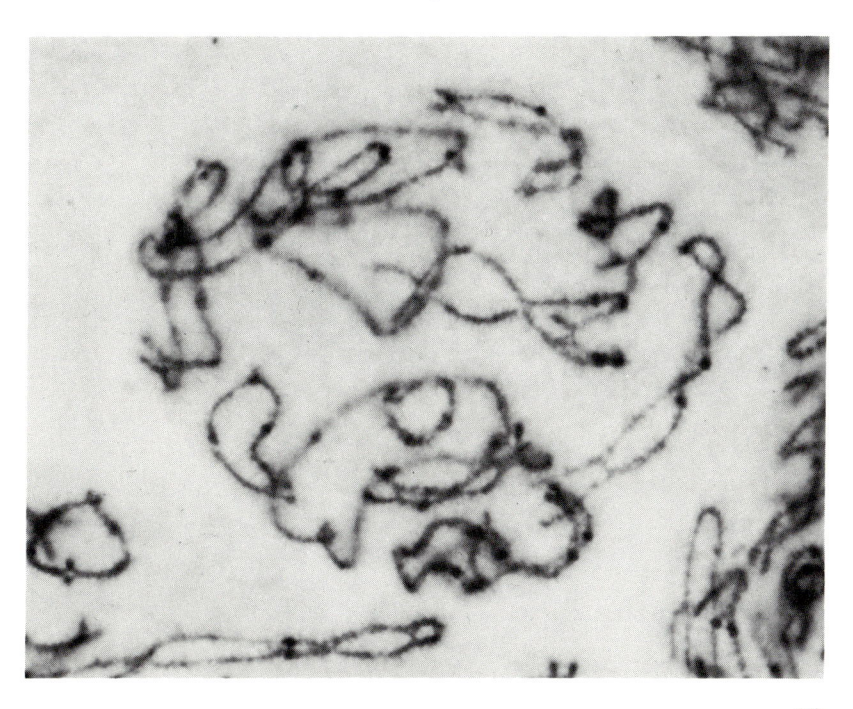

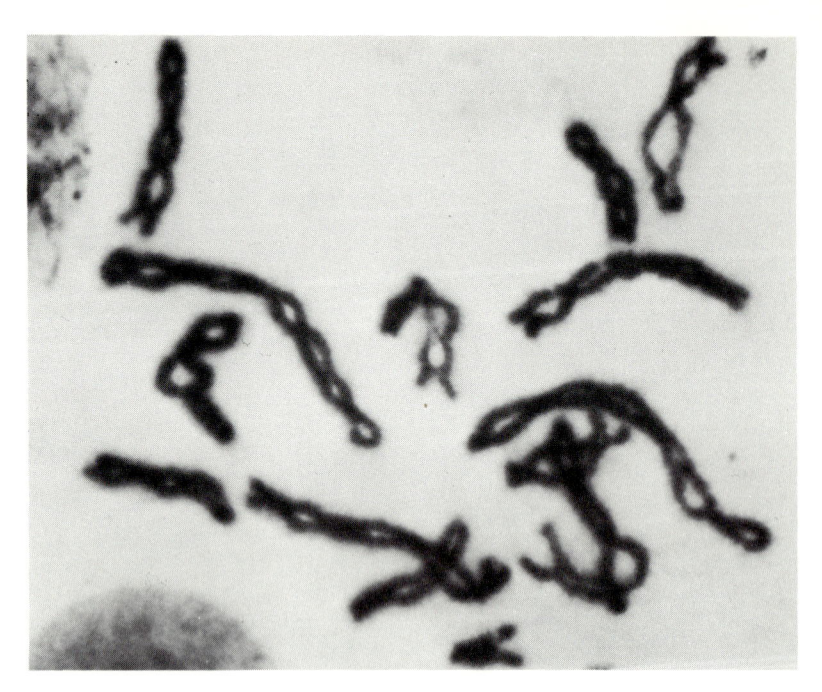

FIG. 6S.9. **Ensatina eschscholtzii oregonensis.** *Late diplotene with high chiasma frequency.*

FIG. 6S.10. **Thorius pennatulus.** *Diplotene in which the four strands of each bivalent are clearly shown.*

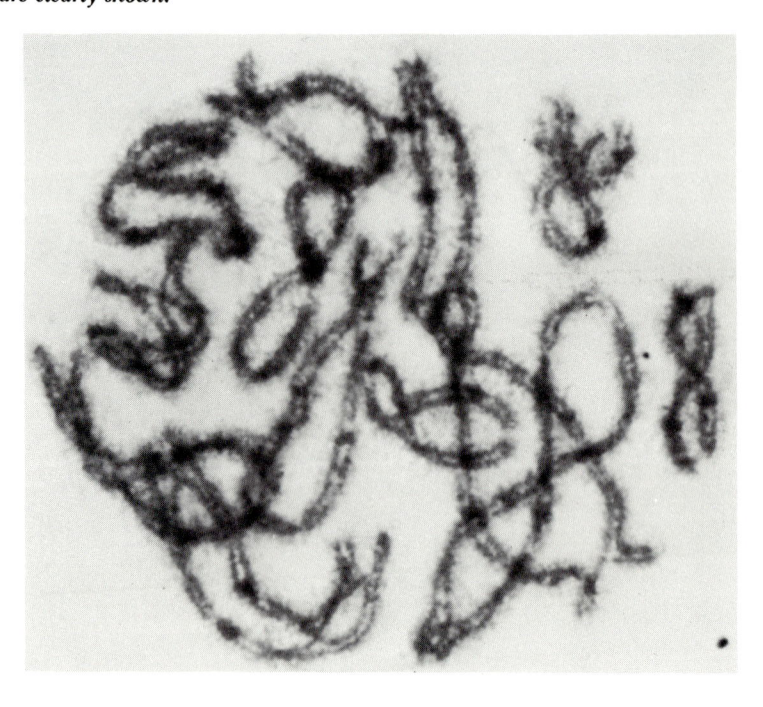

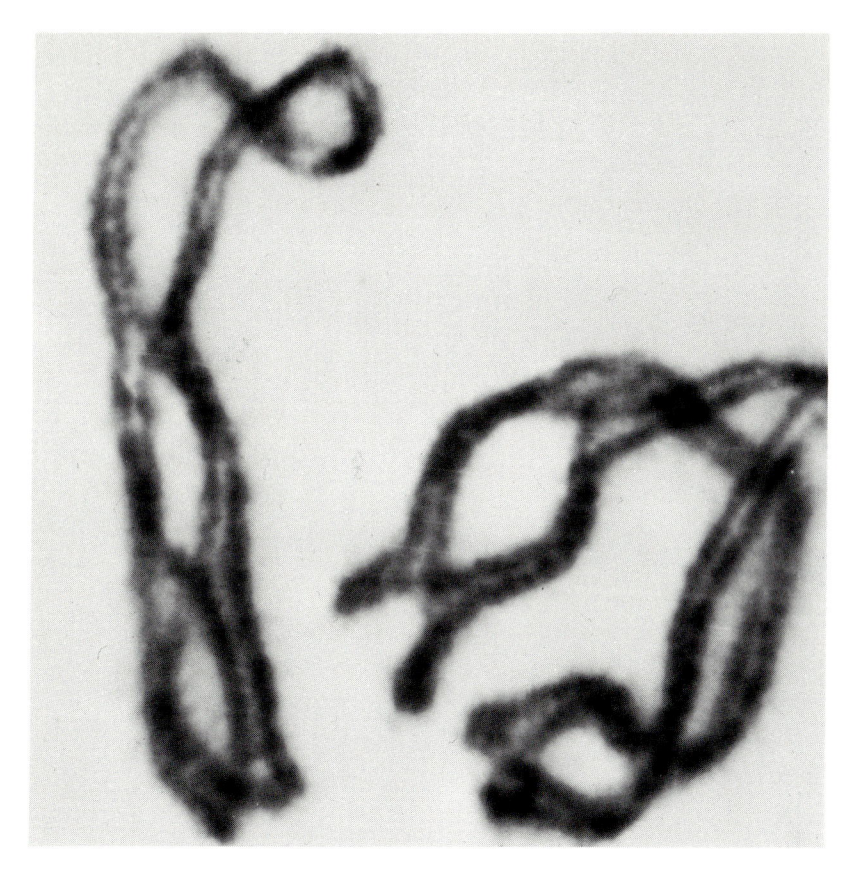

FIG. 6S.11. Batrachoseps wrighti. *Diplotene bivalents.*

FIG. 6S.12. Oedipina poelzi. *Diplotene bivalent. A detailed analysis of this bivalent is presented in the text.*

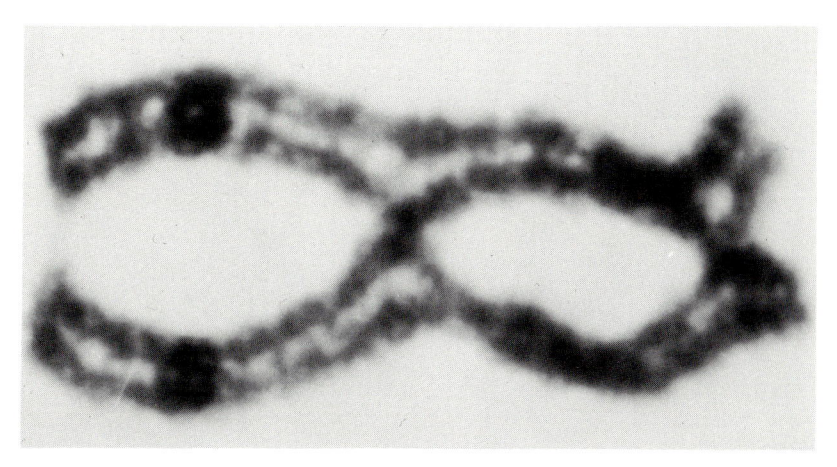

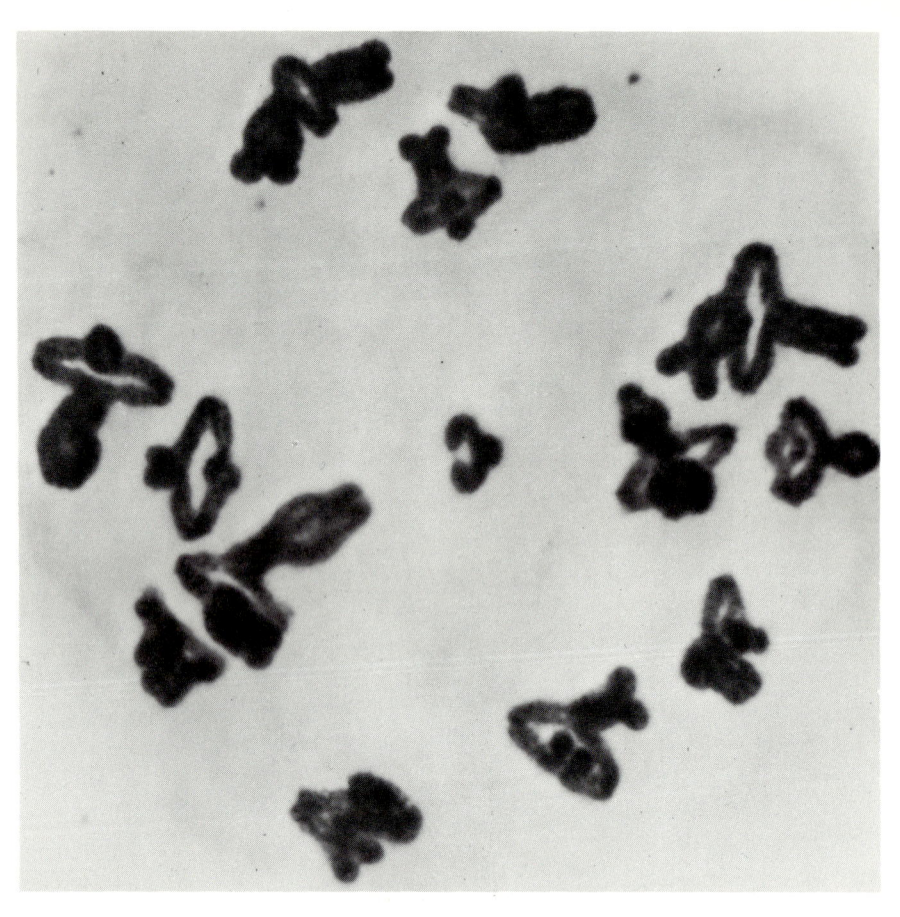

FIG. 6S.13. **Plethodon jordani jordani.** *First meiotic metaphase, polar view.*

FIG. 6S.14. **Plethodon jordani metcalfi.** *First meiotic metaphase, equatorial view.*

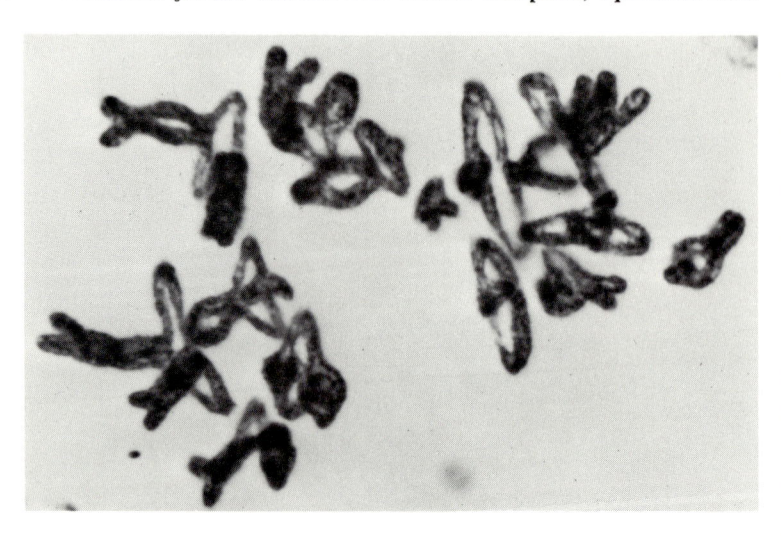

FIG. 6S.15. **Batrachoseps wrighti.** *Early anaphase of the first meiotic division.*

FIG. 6S.16. **Oedipina uniformis.** *Late anaphase of the first meiotic division.*

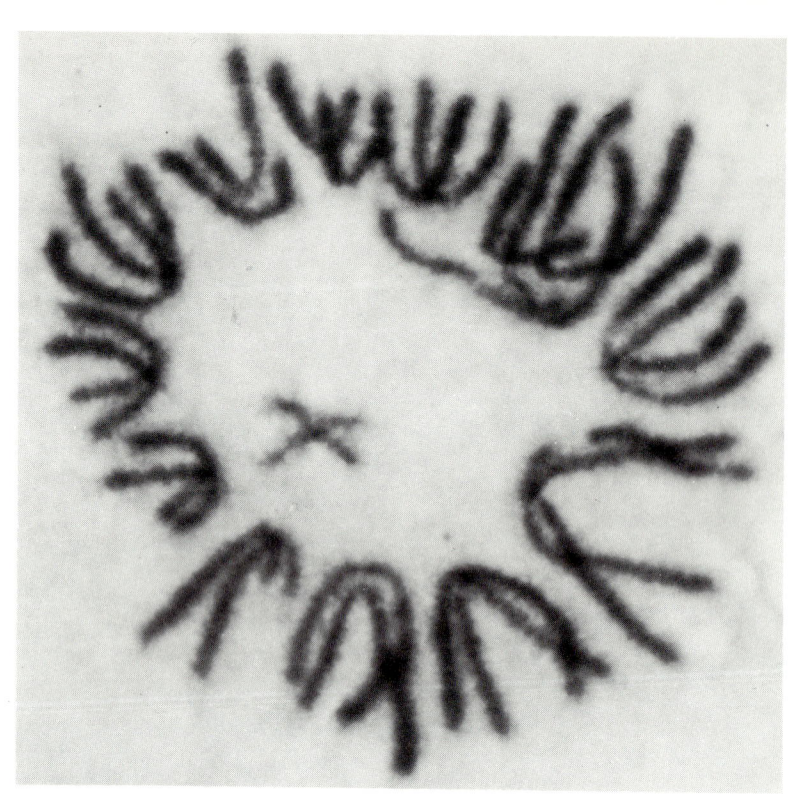

FIG. 6S.17. Ensatina eschscholtzii oregonensis. *Second meiotic metaphase, polar view.*

FIG. 6S.18. Hydromantes shastae. *Second meiotic metaphase, equatorial view.*

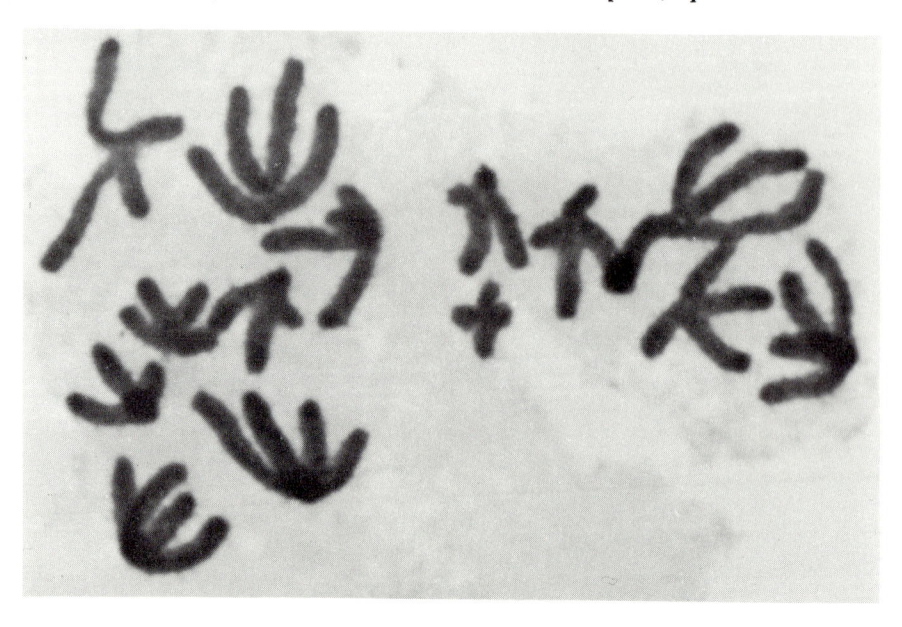

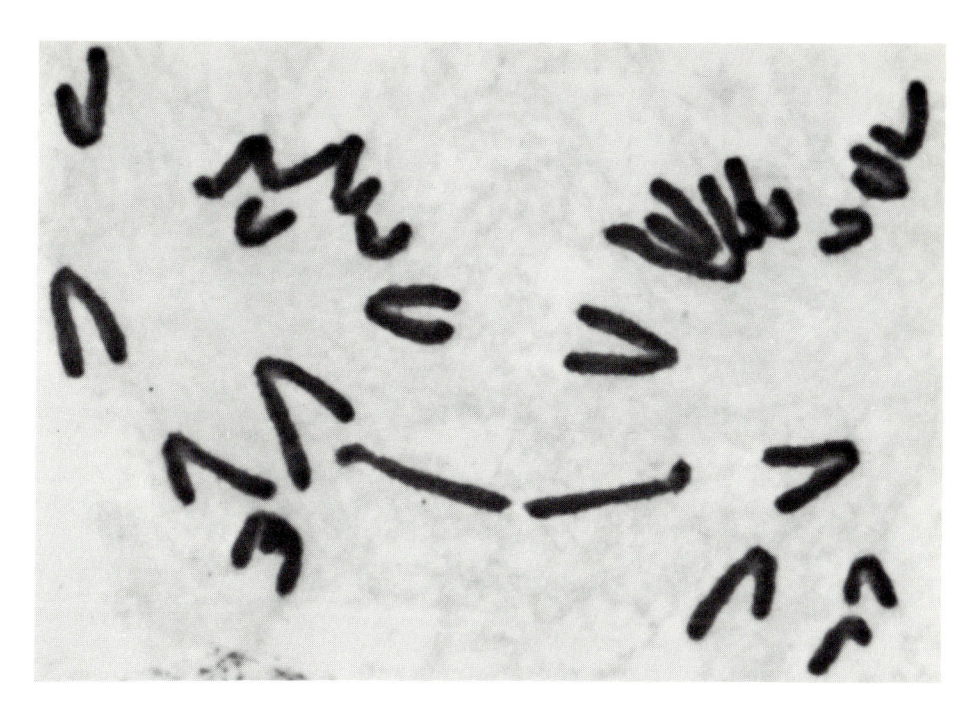

FIG. 6S.19. **Batrachoseps wrighti.** *Second meiotic anaphase.*

FIG. 6S.20. **Batrachoseps wrighti.** *Late anaphase of the second meiotic division.*

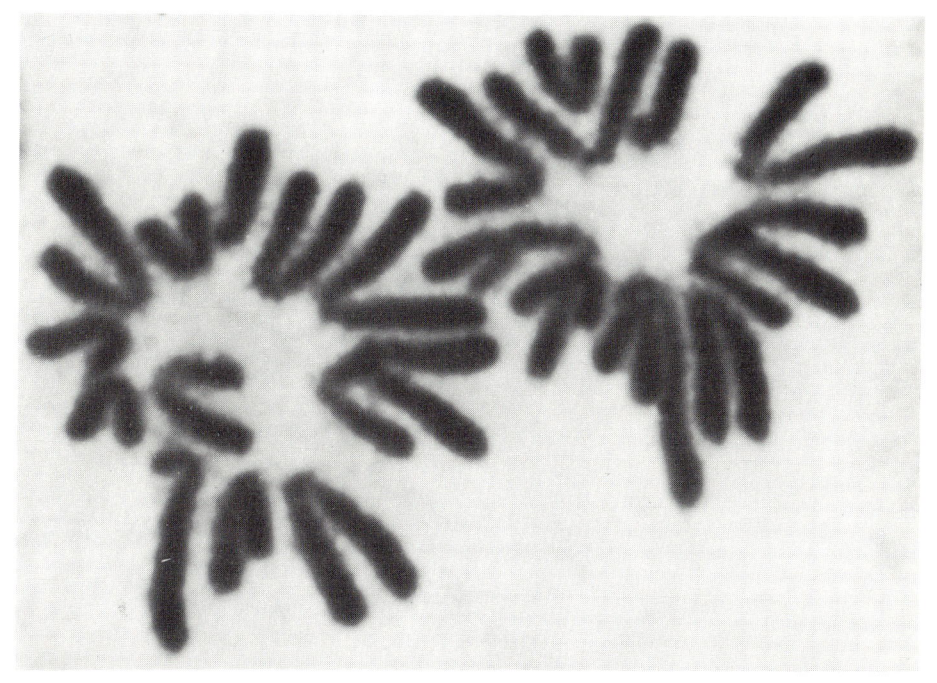

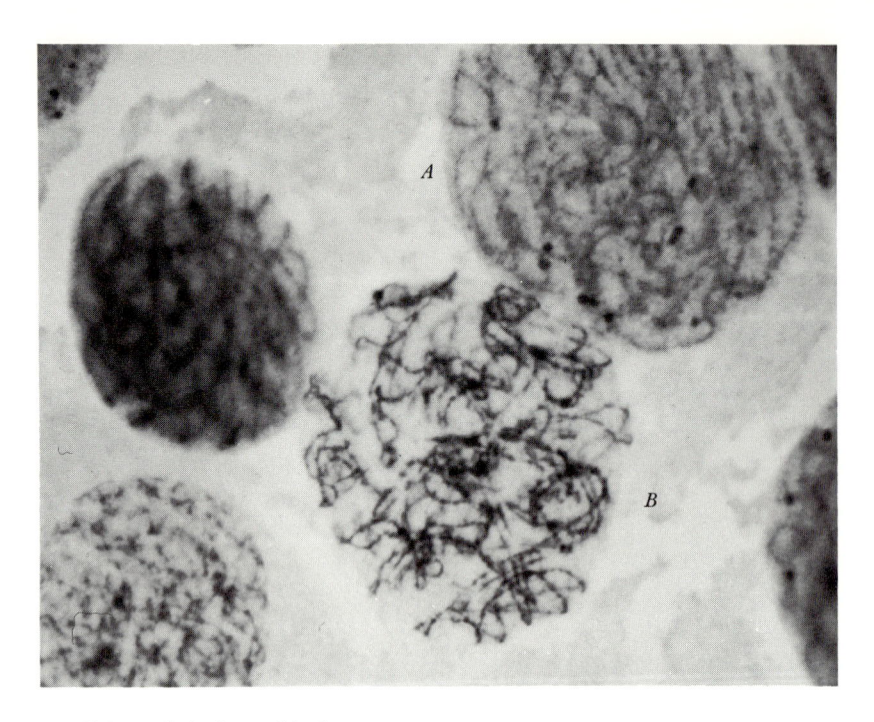

FIG. 6S.21. **Plethodon vehiculum.**

FIG. 6S.22. **Plethodon vehiculum.**

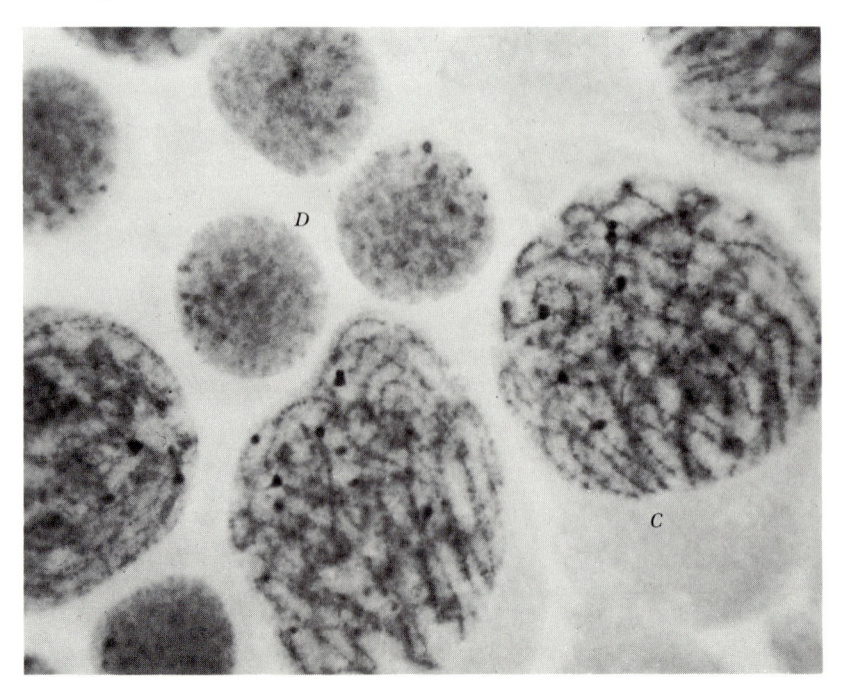

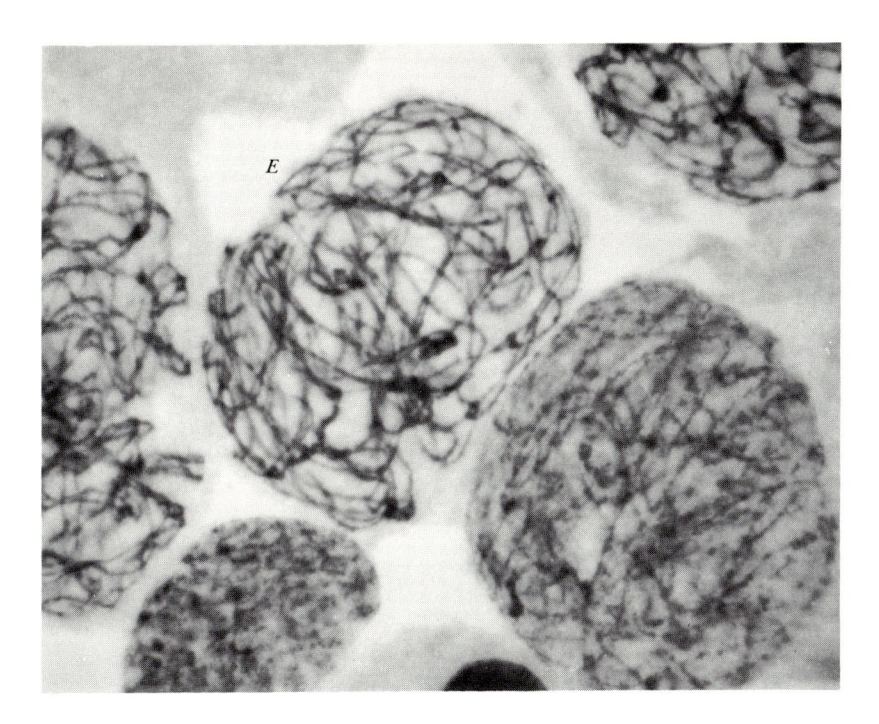

FIG. 6S.23. **Plethodon vehiculum.**

FIG. 6S.24. **Batrachoseps major.**

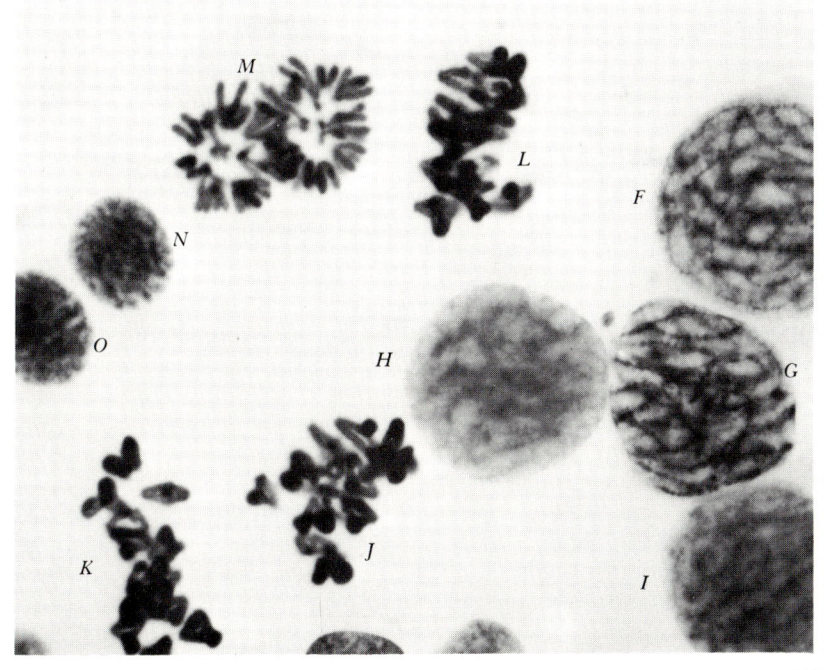

125

one must have cells from a meiotically active testis. The two meiotic divisions appear in the spermatocytes of temperate zone salamanders once each year; they may occur in the spring, summer, or fall, depending upon the species of salamander. The Chromosome preparations are made by a widely used technique that will be briefly described. A suspension of spermatocytes, in which meiotic divisions are occurring, is made on a slide and mixed with a drop of the stain-fixative aceto-orcein (2 percent orcein in 45 percent acetic acid). A cover glass is placed over the suspension of stained cells, the slide is inverted over absorbent paper, and the cells are squashed by pressure on the back of the slide. This squashing separates the Chromosomes and pushes them into a single plane so that their observation under the microscope is greatly simplified. A good grade of orcein will stain only the Chromosomes; cytoplasmic structures, such as those seen in Fig. 1, are not visible in the aceto-orcein squashes. In observing the photomicrographs, remember that the normal arrangement of the chromosomes within the cell has been disrupted by the squashing.

The Chromosomes first become visible as faintly staining, elongated strands in the leptotene stage of the first meiotic prophase (Fig. 6S.2). Chemical studies indicate that DNA duplication occurs in the interphase prior to leptotene, but the double nature of the genic material is not visible under the microscope until later in the first meiotic prophase. The synapsis of the pairs of homologous Chromosomes begins during zygotene, illustrated in Fig. 6S.3. Here and there in these nuclei it is possible to see that the strands are arranged in pairs, indicating that the Chromosomes that appeared to be scattered at random in the premeiotic nucleus are now becoming intimately associated as pairs of synapsed homologues. In zygotene, the elongated strands of leptotene become shorter and thicker and they appear as loops with the free ends directed approximately toward the part of the cytoplasm in which the centrosome is located. The synapsed pairs of homologous Chromosomes are called bivalents.

Pachytene nuclei are shown in Fig. 6S.4. In other organisms, for example, maize, the Chromosomal duplication that took place prior to leptotene becomes visible at this stage so that the pachytene loops appear as four-strand structures. But in aceto-orcein squashes of salamander spermatocytes, the individual strands of the synapsed homologues are so closely associated at pachytene that they cannot be clearly resolved. A typical salamander spermatocyte pachytene nucleus contains the haploid number of bivalent loops, oriented within the nucleus in a manner similar to that seen in zygotene.

In salamanders, the preleptotene Chromosomal duplication becomes visible during diplotene. Figs. 6S.5 and 6S.6 illustrate very early diplotene nuclei in which the bivalent halves are just beginning to separate, but they have not separated sufficiently to reveal the location of the chiasmata. The magnification of these photomicrographs is not great enough to show the four-strand structure of the early diplotene bivalents. With greater magnification and excellent resolution, the four strands making up each bivalent would be evident in these nuclei. Figures 6S.5 and 6S.6 have been selected to provide a comparison of early diplotene nuclei with nuclei at pachytene and zygotene. As diplotene progresses, the bivalent halves become increasingly separated except at the chiasmata, where pairs of homologous strands are in intimate contact (Figs. 6S.7, 6S.8, and 6S.9). It is at these locations that the breaks and exchanges of crossover are believed to occur. Thus, unless appearances are deceiving, the chiasmata represent the visible expression of genic recombination of homologous strands. The four-strand structure of diplotene bivalents is shown with remarkable clarity in Figs. 6S.10, 6S.11, and 6S.12. Salamander spermatocyte squashes only rarely result in preparations showing the superb detail of these photomicrographs.

Let's consider the four strands making up the diplotene bivalents. Duplication in the interphase preceding the first meiotic prophase gives rise to Chromosomes consisting of two strands that are genically identical, assuming no mistakes occur in the duplication process. Identical strands produced by duplication are called sister strands; thus each Chromosome enters the first meiotic prophase as two closely associated sister strands. Since a bivalent consists of a pair of synapsed homologues, it follows that two of the four strands of a particular bivalent are homologous to the other two and, having been derived from different individuals, may be genically different. These relationships can be illustrated by utilizing the bivalent shown in Fig. 6S.12. The strands of this bivalent appear to be associated as shown in the accompanying diagram. Note that only two of the four strands are involved at a particular point of chiasma and also note that the chiasmata occur between homologous strands, not between sister strands.

Metaphase of the first meiotic division is shown in Figs. 6S.13 and

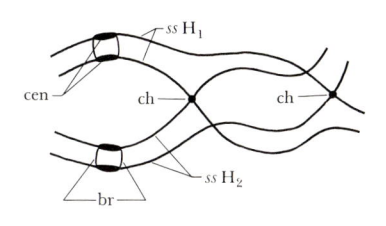

$ss\ H_1$ and $ss\ H_2$ = sister strands of the two homologues

ch = chiasma

cen = centromere. There are four of these, one for each strand.

br = bridging material holding the centromeres of sister strands together.

6S.14. The bivalents have moved to the equatorial position within the cell and the homologous centromeres of the bivalent halves have become widely separated. The four strands of each bivalent are dismembered by the two meiotic anaphases. The anaphase of the first meiotic division separates the four strands two-from-two, and the second meiotic anaphase separates the resulting two-stranded half-bivalents one-from-one. In Fig. 6S.15 the anaphase of the first meiotic division has just begun: the homologous centromeres are moving in opposite directions, bringing about a two-from-two separation of the four strands of each bivalent. The two strands making up each of the half-bivalents remain attached to each other by means of material in the regions on either side of their centromeres. In Fig. 6S.16, the first meiotic anaphase has been completed.

The brief interphase between the first and second meiotic divisions is not illustrated in the photomicrographs. There is no Chromosomal duplication during this interphase and it is because of this that the two meiotic divisions succeed in reducing the Chromosome number by one-half. If a cellular division occurs without a preceding Chromosomal duplication, the Chromosome number will inevitably be halved, assuming that half of the Chromosomes go into one of the resulting two cells and the remaining half into the other cell. In meiosis, there are two cellular divisions with only a single duplication of the Chromosomes and consequently a halving of the Chromosome number in the four resulting nuclei. Thus, in addition to providing for genetic recombination, the meiotic divisions constitute the mechanism whereby a diploid organism produces gametes with only half the number of Chromosomes of somatic cells. This halving of the number of Chromosomes is clearly a necessary complement to the doubling of Chromosome number resulting from the union of gamete nuclei at fertilization.

Metaphase of the second meiotic division is shown in Figs. 6S.17 and 18, in which the half-bivalents are seen at the equatorial position. A one-from-one separation of these two-strand half-bivalents occurs at the anaphase of the second meiotic division, as shown in Figs. 6S.19 and 6S.20. These two photomicrographs were obtained from a salamander in which the diploid Chromosome number is 26. But there are only 13 Chromosomes in each of the groups shown here, clearly indicating that the meiotic divisions have brought about a halving of the Chromosome number. Note particularly that the pairs of homologous Chromosomes that existed together in the spermatocyte during prophase and metaphase of the first meiotic division (and that became genically modified by recombination), are separated by the two meiotic divisions so that each of the four cellular products of meiosis contains only one member of particular homologous pair.

Figures 6S.21, 6S.22, 6S.23, and 6S.24 have been included to enable the student to test himself on his understanding of meiosis. Some of the nuclei of these photomicrographs have been designated by letters. Try to identify the meiotic stages represented by these nuclei and then check your decisions with the information presented in the following paragraphs.

Before checking your answers, we should face the fact that it is frequently difficult to fit a particular nucleus precisely into one of the divisions of the first meiotic prophase. In part, this is because the process is continuous and thus gives rise to intergrade nuclei that are not easily forced into the traditional categories. Classification is also complicated by the fact that some of the events by which the stages are defined are difficult to resolve even with the best optical equipment. In Fig. 6S.21, nucleus A is probably best classified as zygotene, although the synapsis of homologous strands is not as clearly shown as might be desired. Nucleus B is clearly in early diplotene: the halves of the bivalents have separated and some of the positions of chiasmata can be seen.

In Fig. 6S.22, nucleus C is again an example of zygotene. Note how the strands are bent into loops with their free ends gathered together in the lower portion of the nucleus. A clue as to the position of nucleus D in the meiotic process is provided by its size. There are two possibilities here: A nucleus of this size could be derived from a cell that has just completed the spermatogonial mitoses and is now ready for the increase in size that will bring it to the dimensions of a leptotene spermatocyte. Alternatively, it could be a nucleus in the interphase between the first and second meiotic divisions, its relatively small size reflecting the fact that the first meiotic division has been completed. The former of these two possibilities is the more probable, since there were no first meiotic stages later than zygotene in the part of the slide from which this photomicrograph was taken. This illustrates how one's decision as to the classification of a particular nucleus may be influenced by the stages of associated nuclei, since there is a high degree of synchronization of meiotic stages among the cells of a testis lobule.

Nucleus E of Fig. 6S.23 is an excellent example of very early diplotene. Fig. 6S.24 is particularly interesting because of the many different stages that are displayed in a single field. Nuclei F, G, H, and I are in early prophase of the first meiotic division and they form an intergrading series from leptotene to pachytene. The strands of H are sufficiently indistinct to justify a classification of leptotene; I and probably also F are zygotene, but F seems to be approaching pachytene and, at the magnification of the photomicrograph, precise identification is difficult. Nucleus G seems to fulfill the criteria for pachytene. Nuclei J, K, and L are at the first meiotic metaphase, M is an anaphase of the first meiotic division and N and O are nuclei in the interphase between the first and second meiotic divisions.

Summary

In conclusion, let us think back over the events of meiosis and summarize them. (1) Chemical studies show that a Chromosomal duplication takes place during the interphase prior to the prophase of the first meiotic division. This doubleness of the Chromosome strands does not become visible until later. (2) Synapsis occurs during the zygotene stage of the first meiotic prophase, bringing the homologous strands into an intimate contact. (3) The duplication that took place during the previous interphase becomes visible during diplotene, giving rise to the four-strand bivalents. (4) As the diplotene stage proceeds, the chiasmata among the homologous strands become clearly indicated. These are believed to be the points at which breaks and exchanges take place; thus, it is at this time that recombination of the genic material of homologous strands apparently occurs. (5) The four strands of each bivalent are dismembered by the two meiotic anaphases. Homologous centromeres are separated at the first meiotic anaphase and the sister centromeres of the resulting half-bivalents are separated at the second meiotic anaphase. (6) The four cells that result from the two meiotic divisions contain only half the number of Chromosomes of the cell from which they have been derived, since the two cellular divisions involve only a single Chromosomal duplication. (7) The two members of each homologous pair of Chromosomes, genically modified by recombination, are separated from each other by meiosis and segregated into different nuclei so that each of the four products of a meiosis will have only one member of a given homologous pair. Random nuclear fusions among these haploid cells (the sperms and eggs of salamanders) will restore the diploid Chromosome number, bringing the modified pairs of homologues into the intimate association of a single nucleus and making possible still other kinds of genic combinations during the succeeding meiosis.

Seven

Recombination in Viruses

Because the DNA-phages are the only viruses upon which significant recombination studies have been conducted, we shall confine our discussion to them.

A Phage Cross

A "cross" in the case of phages occurs as a consequence of simultaneous infection of a cell by two or more different types of the same phage. (There are no mating types among the phages.) Such mixed multiple infections are accomplished in the laboratory as follows: bacteria (at high concentration, say 10^8 cells per ml, to promote adsorption of phages) are added to a suspension of two phage types, each at a final concentration several times that of the bacterial concentration. For simplicity, let us suppose that the two types differ at only two loci, and for brevity let us call the two infecting types $a+b$ and $ab+$. Four types of phages are observable among the particles which mature earliest (see Problem 3.3)—the infecting "parental" types and the recombinant types ab and $a+b+$. Populations of particles maturing in successive intervals contain progressively higher frequencies of recombinants.

The dependence of recombinant frequency on time of maturation suggests that we must consider the multiplying pool of chromosomes as a population whose members have repeated opportunities for genetic exchange. This notion is strengthened by the results of triparental crosses. If bacteria are infected by three types of particles, say $abc+$, $ab+c$, and $a+bc$, *eight* kinds of particles are found among the offspring. Since one of these eight classes ($a+b+c+$) carries markers from all three infecting types, we must again envision successive, or otherwise complex, acts of exchange. These complexities, however, do not prevent us from constructing linkage maps.

For all the well-studied phages, all the marked loci of the phage are linked to each other. (This result is in accord with our observation in Chap. 5 that a virus particle contains a single nucleic acid molecule.) For several different phages the map order of a number of markers has been determined from the results of crosses performed as described above. These results are characterized by two features: (1) the frequency of recombinants observed for the most loosely linked markers is measurably less than 0.5; (2) coefficients of coincidence (determined in either two- or three-factor crosses) are greater than unity. Is there a simple meaning to these two properties of phage crosses? In general terms we can propose heterogeneity among the emerging particles with respect to their opportunities to have become recombinant. Thus, some of the particles had no such opportunities and necessarily contribute to the frequency of parental-type particles. Those particles which did have opportunities to recombine may have done so in several places along the chromosome. Thus, one observes a high frequency of parental types and a positive correlation ("negative interference") for recombination in two regions of the linkage map. Two sources of heterogeneity among phage particles emerging from a cross can be securely identified; others may exist. (1) In a standard phage cross, several (say 5 to 10) particles of each parental type are adsorbed per bacterium. Not all the bacteria adsorb the same number of either type of particle. In fact, the particles are more or less Poisson-distributed among the bacteria. Thus, the ratio of the infecting, parental types varies from cell to cell. In cells with a preponderance of one type a greater number of the exchanges which underlie recombinant production occur between chromosomes of like type. Particles emerging from these cells thereby have a less than average chance of being recombinant. (2) Mature phage particles form over a period of time (see Problem 3.3). Those chromosomes which at a very early phase of the process have been removed by maturation from the recombining population have had fewer opportunities for recombination than those which are removed later.

Despite the complexities described above, the genetic markers of phage can be ordered. Those markers giving the largest recombinant frequencies in crosses with each other are defined as the farthest apart.

Comparison of a Phage Map with Its Chromosome

The distances between markers on the chromosome are in principle measurable in conventional physical units (for instance, microns) or chemical units (numbers of base pairs). Either metric is additive; for markers *a, b,* and *c* arranged in that order, the distance between *a* and *c* is equal to the distance between *a* and *b* plus the distance between *b* and *c*. In order to make a quantitative comparison between a linkage map and a chromosome, we must construct the linkage map such that the distances are, like physical distances, additive.

In Chap. 6 we looked at the function which relates map distance to recombinant frequency for systems in which there is no interference ($S = 1$ at all values of p). We used the function to illustrate, with a priori assumptions, the nature of a relationship between the mean number of crossovers between two loci (which is proportional to the "map distance") and the recombinant frequency. As implied in Chap. 6, however, the "ideal" mapping function rarely holds; one typically observes a set of p values for which $S \neq 1$ and, in fact, for which S varies as a function of p. In order to construct a proper map, therefore, one must divine some other mapping function to convert recombinant frequencies to distances which are additive.

Several approaches have been employed for constructing mapping functions. In the case of phages, one of these approaches has been to attempt to relate map distances to recombinant frequencies by formalizing an hypothesis regarding the goings-on in an infected cell. Most such approaches have been based on the mating theory of N. Visconti and M. Delbrück. These approaches can "work" (that is, they can convert recombinant frequencies to additive distances), but the formalizations are too involved to present here. Furthermore, the highly questionable assumptions which go into them are more likely to mislead than to enlighten beginning students (or most anyone, for that matter). A more generally applicable approach to the problem, described below, has been applied to the phage λ.

Consider a set of three loci linked in the order 1–2–3. Let R_{12} and R_{23} be the recombinant frequencies observed in crosses involving loci 1 and 2 and 2 and 3 respectively. Then, by analogy with Eq. 6.5, we can write

$$i = \frac{\text{Frequency of double recombinants}}{R_{12}\ R_{23}} \quad \text{(Eq. 7.1)}\ [1]$$

[1] We introduce symbolism here which is different from that used in the previous chapter in order to modulate the temptation to think of viral recombination in a meiotic framework.

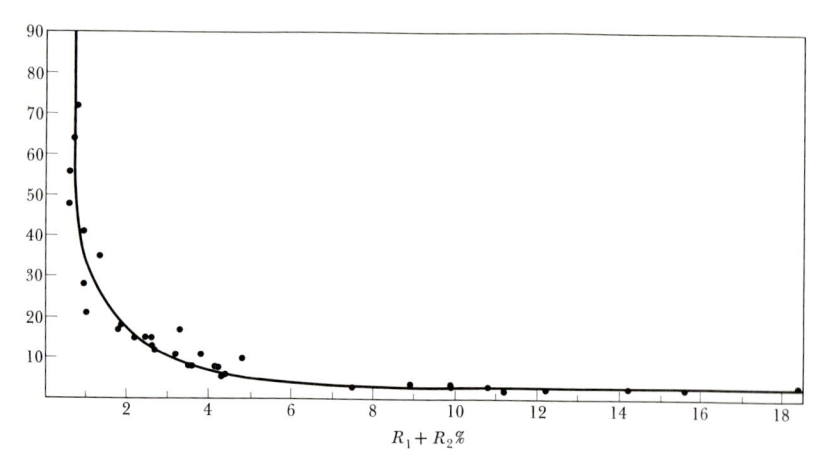

FIG. 7.1. *The relationship of the coefficient of coincidence* i *to the sum of the component recombinant frequencies in each of 36 different three-factor crosses in phage* λ. *From P. Amati and M. Meselson,* Genetics, 51 *(1965), 369–79.*

Values of i (coefficient of coincidence) determined in three-factor crosses were found to vary with R as shown in Fig. 7.1. The salient features of these data are: (1) i is greater than unity for all values of $R_{12} + R_{23}$. For the larger values of R this presumably reflects the population heterogeneities described above; (2) i rises sharply for small values of $R_{12} + R_{23}$. We shall return later to the *biological* significance of this point; now let's get on with map construction. In the following I shall borrow heavily from Amati and Meselson (see References at the end of this chapter) without confusing you with quotation marks.

The data of Fig. 7.1 may be used to construct a function which transforms recombinant frequencies into additive distances. That is, we may obtain a function giving the number of unit intervals each with recombinant frequency R_0 which may be fitted into any larger interval having frequency R. We choose as unit interval (for this case of phage λ) $R_0 = 0.01$ percent because frequencies of this magnitude or less will be almost perfectly additive if the coefficient of coincidence i never rises much above the greatest observed value of approximately 100. This may be seen from the expression

$$R_{13} = R_{12} + R_{23} - 2iR_{12}R_{23} \qquad \text{(Eq. 7.2)}$$

(which is an analogue of Eq. 6.6), where i is the coefficient of coincidence defined by Eq. 7.1.

To construct our mapping function we assume the existence of a continuous function i^* for the special case $R_{12} = R_{23}$. Then we may use Eq. 7.2 to find the recombinant frequency $R(2)$ for any interval containing two unit intervals:

$R(2) = 2R_0 - 2i^* R_0^2$

Next, we use this value of $R(2)$ to compute $R(4)$, the frequency for any interval containing four unit intervals:

$R(4) = 2R(2) - 2i^* R(2)^2$

In this manner, we may compute $R(2)$, $R(4)$, $R(8)$, and so on, building up a table relating observed frequencies to the number of unit intervals they contain. At each step we can use the appropriate value of i^* from Fig. 7.1 since the inequalities between R_{12} and R_{23} in the various three-factor crosses used to obtain these data are not extreme and were not found to have a great effect on i. The mapping function found by this procedure is shown in Fig. 7.2. By means of the graph we can convert a set of observed recombinant frequencies in λ to map distances which will be, like physical distances, additive. Maps con-

FIG. 7.2. *A mapping function for phage* λ. *The function was constructed from the curve in Fig. 7.1 by the method described in the text. Note that the units "Map Distance" were chosen to be proportional to the observed recombinant frequency in the range where recombinant frequencies themselves are essentially additive. From P. Amati and M. Meselson, Genetics, 51 (1965), 369–79.*

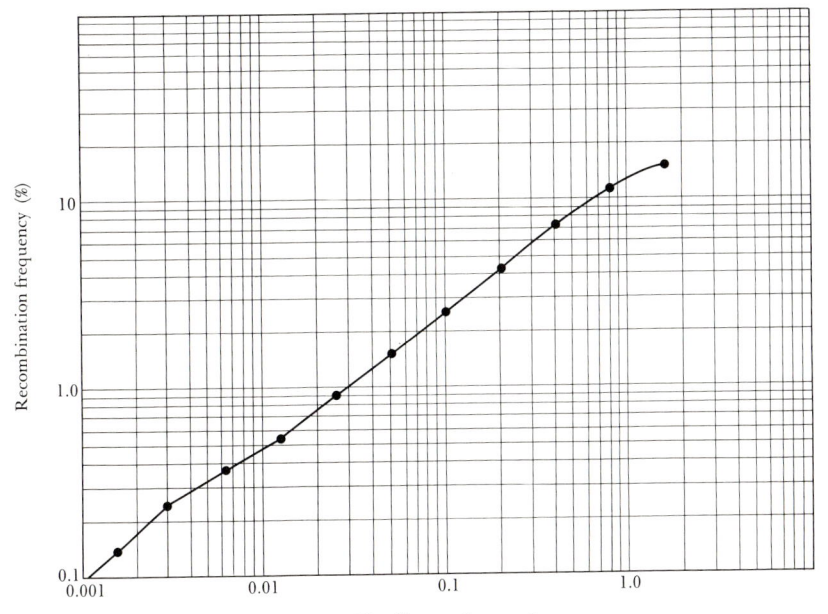

structed so as to locate the positions of markers in such units can be compared with other maps which are derived by methods which are a priori more likely to reflect true physical distances. For phage λ, the correspondence between the "linkage map" and other, "physical maps," is good; the two kinds of distances are approximately proportional by the same factor along most of the chromosome throughout the range investigated.

The relationship between the genetic map of λ and the chromosome of λ was established in the early 1960s by experiments from the laboratories of M. S. Meselson and J. J. Weigle, and D. Kaiser and D. Hogness. Since Meselson and Weigle's experiments have especially relevant overtones, let's look at Kaiser and Hogness' experiments first.

An isolated chromosome of λ can (with rather low efficiency) enter a bacterium and multiply in the presence of coinfecting intact λ. The success of such infections is signaled by the appearance among the progeny phages of genetic markers distinguishing the isolated chromosomes. One of the first rewards of the careful development of this experimental system is derived from its application to the study of "half-chromosomes." When a DNA solution is stirred, the DNA molecules are fragmented. Since the shear forces generated by stirring are maximal near the middle of a molecule and are positively correlated with the length of a molecule, stirring at controlled speeds can produce a population of fragments all of which have approximately equal molecular weight; the break in each chromosome occurs near the middle. When these half-chromosomes infect bacterial cells along with intact phages, progeny phages appear that carry markers from the half-chromosomes. When the phage yield from individual cells is examined, it can be seen that a given half-chromosome contributes markers from *half* of the genetic map. At this first glance, then, map distances seem to bear a pleasant relationship to physical distances. The experiment has also been done with quarters and eighths of chromosomes. The permissible conclusions are limited by the fact that only the two terminal fragments of a λ chromosome can invade an infected cell. (In Chap. 8 we shall see what's special about the ends of the λ chromosome.) Within those limitations the two kinds of maps correspond satisfactorily.

The experiments of Meselson and Weigle show an equally pleasant correspondence between linkage map distance and physical dimensions of the λ chromosome. Meselson and Weigle infected cells with λ particles carrying genetic markers at two loci located like this on the map:

```
          x     x
_____
```

They coinfected with λ heavily substituted with N^{15} and C^{13}. Progeny particles were scored both with respect to their genetic markers (their genotype) and the amount of inherited isotope. Most of the particles of each of the four genotypes contain only DNA synthesized completely

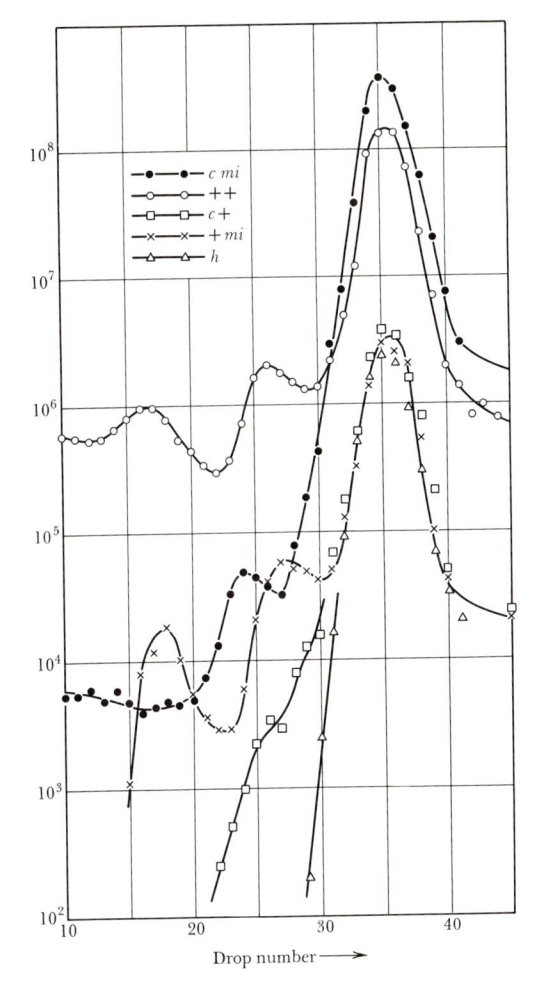

FIG. 7.3. *The transmission of labeled DNA into progeny particles from a genetic cross. Wild type* $\lambda(++)$ *heavily labeled with* N^{15} *and* C^{13} *was crossed with doubly mutant* λ(c mi) *containing the ordinary isotopes* C^{12} *and* N^{14}. *The host cells were grown in ordinary medium. The progeny particles were centrifuged to equilibrium in a CsCl density gradient. Successive drops collected after puncturing the bottom of the centrifuge tube were assayed for the titer of each of the four phage genotypes. The centrifuge tube contained* λh *particles to provide a density and band-shape reference. The significance of the distributions is discussed in the text. This figure, reproduced by permission of M. S. Meselson and J. J. Weigle, appeared in* Proc. Natl. Acad. Sci. U.S., 47 *(1961), 863.*

de novo from the light isotopes of the bacterial culture medium. A small fraction of the particles of each genotype assume positions in an equilibrium density gradient that reflect inheritance of DNA from the isotopically labeled infecting particles. The density distribution of each of the genotypes is depicted in Fig. 7.3.

Detailed interpretations of the distribution of labeled DNA among individuals of each of the four genotypes are still being evolved with the aid of additional experiments; we'll confine our attention here to the relevant features of the distributions.

(1) In addition to the density mode corresponding to completely unlabeled particles, two other modes for the labeled parental genotype are apparent. One of these (drop 17 in Fig. 7.3 reflects the

occurrence of progeny phages that have inherited a complete, labeled DNA molecule.[2] The second mode (drop 26) reflects particles that have inherited a semiconservatively replicated chromosome.

(2) The recombinant type ($+mi$) that has inherited the "left-hand" marker on the map of the labeled parent also shows two density modes reflecting substantial inheritance of labeled DNA. The more dense mode (drop 18) signals the inheritance of a chromosome labeled in about 90 percent of its nucleotides. The chromosomes in such particles must be labeled over most of their length in both chains. The less dense mode (drop 27) signals the inheritance of a chromosome labeled in about 45 percent of its nucleotides. The two density modes probably result from particles containing chromosomes with label distributions like this

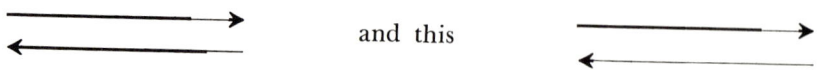

and this

respectively.

For the moment let's be content with drawing the obvious conclusion that recombination can result from the transverse breakage of an unduplicated chromosome.

(3) The recombinant type ($c+$) that has inherited the "right-hand" marker of the labeled parent has a density distribution indicating that the inheritance of that marker from the labeled parent does not involve a conspicuous inheritance of labeled DNA.

The density distributions reveal that the two recombinant genotypes share unequally in the inheritance of labeled DNA. The asymmetry in isotope inheritance correlates well with the asymmetric distribution of the two markers on the map.

The map of phage λ, as constructed from data collected in standard crosses, has two ends—and the two ends of the map correspond with the two ends of the λ chromosome. Phage T1 is similar to λ in this respect. However, analyses comparable to those carried out for λ reveal that the linkage map of T1 is stretched at the ends relative to the physical map; the probability of recombination per nucleotide is higher near the ends than near the middle. The chemical basis for this variation in recombination frequency along the chromosome has not yet been established.

Many phages differ from λ and T1 in that their linkage maps are circular. For each of these phages analysis of the structure of the chromosome has revealed the physical basis of the linkage map circularity. Some of the phages with circular maps have circular chromosomes. Others of them have linear chromosomes in which the nucleotide

---

[2] These progeny could conceivably be derived from DNA that has duplicated only conservatively. It seems more likely, however, that these fully labeled chromosomes never did duplicate under the experimental conditions of multiple infection.

sequences of the chromosomes in a clone of particles are circular permutations of each other. We shall take a closer look at these permuted chromosomes later.

In Fig. 7.1 we saw that the coefficient of coincidence rose at small values of R. The origin of this "high negative interference" is obscure, but is the object of considerable research. Two classes of explanations are under investigation. Members of the first class are formal explanations not closely hinged to our knowledge of DNA structure. Members of the other class depend intimately upon our knowledge of DNA structure. We shall consider the first class below and the second class in the next section. The two classes are not mutually exclusive.

The first class of models for the origin of high negative interference supposes that the conditions which permit or promote crossing over happen only to short stretches of the chromosome. One version imagines that the chromosomes synapse over a rather short stretch (or stretches) equal in length to the distance over which one sees high negative interference. Within this stretch several crossovers can then occur. Another version imagines that the enzyme(s) which cut and rejoin the chromosomes are apt to strike several times in the same neighborhood simply because they are there. Neither of these models specifies the mechanism of crossing over—they merely construct excuses for a spatial clustering of them.

The "Mechanism" of Recombination

A second phenomenon which is correlated with recombination between close markers forces our consideration of crossing over in terms of DNA structure. A small fraction of the particles emerging from a phage cross give rise to progeny particles of two different types. These particles (at least the ones we are talking about now) are ordinary in that they contain only one chromosome but are extraordinary in that they carry two allelic markers! The frequency of these "heterozygotes" is only 1 percent or less in the overall population under ordinary conditions but is many times higher among particles recombinant for two closely linked markers. This correlation led to the idea that crossing over involves the union of two pieces of (duplex) DNA into a structure like this:

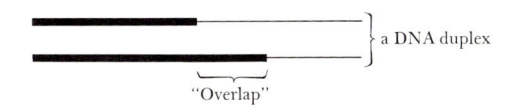

$$\underbrace{\rule{4cm}{0pt}}_{\text{"Overlap"}} \Bigg\} \text{a DNA duplex}$$

where the heavy and light lines indicate the contributions from the respective parent molecules. For markers that are close together compared to the mean length of the overlap, these structures are likely to be recombinant on only one chain. For example, in the cross $a+b \times ab+$ the following recombinants can arise:

$$
\begin{array}{cc}
a^+ & b^+ \\
a^+ & b \\
\\
a & b^+ \\
a^+ & b^+ \\
\\
a & b \\
a & b^+ \\
\\
a^+ & b \\
a & b \\
\end{array}
$$

Each of these recombinants is a heteroduplex for one or the other of the two loci involved. Support for the idea of a heteroduplex basis of phage heterozygosity has been obtained for several phages.

In ordinary phage crosses, the frequency of heteroduplexes among particles matured in successive intervals is roughly constant. Within the framework of the proposed model this constancy implies that the rate of formation of heteroduplexes by crossing over is equal to the rate of their destruction by Watson–Crick duplication (or their dilution by Jehle duplication; see Problem 4.2). If DNA synthesis is depressed during a phage cross, the frequency of heterozygotes rises; upon restoration of synthesis the frequency falls again. This evidence, coupled with the significant failure to find any gross structural abnormality accompanying this kind of phage heterozygosity, constitutes some of the support for our picture of a newly arisen recombinant molecule. Other support comes from physical studies on intermediates in the formation of recombinant molecules. J. I. Tomizawa infected bacteria simultaneously with density-labeled and with radioactivity-labeled T4 particles. Soon after the infection, which was carried out under conditions which block DNA duplication, DNA molecules could be recovered from the cells which were radioactive and which had intermediate density. The mode of attachment of the density and the radioactivity labels to each other was revealed by a variety of tests all of which supported the picture that the recovered molecules contained some long stretches of fully radioactive DNA and some long stretches of fully dense DNA. When these molecules were recovered early after infection, the radioactive and the dense stretches could be separated from each other by conditions which disrupt the hydrogen bonding of short DNA duplexes. For the molecules recovered somewhat later, these conditions failed to separate the two kinds of fragments. Lateral scissions of the duplex introduced by stirring were now required to effect separation of radioactivity label from density label. This and yet other evidence nicely support the idea that the overlap model for recombinant molecules is essentially correct and that an intermediate in the formation of recombinant molecules looks like this:

The two pieces are held together only by hydrogen-bonding in the overlap region prior to the enzymatic sealing of the two chains.

Armed with a picture for the basic structure of a recombinant molecule we can again inquire into the origin of high negative interference. One notion will sound familiar; high negative interference is due to closely spaced occurrences of the basic event so that a recombinant molecule, for example, *really* looks like this:

The other notion imagines that the simple structure is first formed and is then acted upon by enzymes which object to violations of the Watson–Crick rules of base-pairing. Thus, in a cross of (for example) $a+bc+ \times ab+c$, the following structure is one of those which could arise by crossing over:

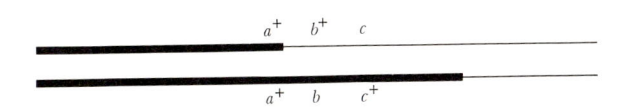

$$a^+ \quad b^+ \quad c$$
$$a^+ \quad b \quad c^+$$

There are two points of base mismatch in the structure (at loci b and c) either of which is supposed to be susceptible to the action of enzymes which remove short stretches from one or the other chain. Following action of these enzymes, the molecules could look like this:

$$a^+ \quad b^+$$
$$a^+ \quad b \quad c^+$$

if, perchance, the hypothesized enzymes removed a stretch of nucleotides including the c marker. A second set of enzymes then comes into play to rebuild the missing stretch. One component of this set could be the DNA polymerase described by Kornberg. The reconstructed recombinant molecule would then look like this:

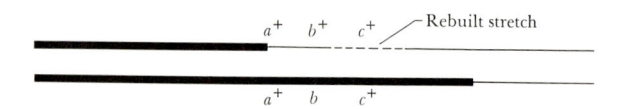

$$a^+ \quad b^+ \quad c^+ \quad \text{Rebuilt stretch}$$
$$a^+ \quad b \quad c^+$$

In this example one of the two chains is recombinant simultaneously in the small intervals between loci a and b and b and c. A subsequent act of duplication or further removal and resynthesis could result in the production of "pure" double-recombinant molecules. This model for high negative interference has an important implication. It implies that the *markers themselves* may influence the events leading to genetic recombination. Our previous discussions took the markers to be merely what that term implies—tags that permit us to observe exchange events which would be going on in precisely the same way in their absence. This assumption applies sufficiently well in enough situations to make it a very useful one. However, be hereby warned that "it ain't necessarily so" in all cases.

The idea of "repair enzymes" acting to "cure" heteroduplex structures derives primarily from studies on radiation-damaged or chemically-damaged phage or bacterial DNA. We shall not attempt to detail these studies. Important for our purposes is that they clearly demonstrate that there are enzymes present in the cells which *can* recognize *some* kinds of aberrations in DNA structure and repair them by removal and resynthesis. In Chap. 9 we shall refer to some of these ideas in our discussion of explanations of certain phenomena associated with recombination between tightly linked markers in higher organisms.

In recent years, the discovery of mutant phages (as well as bacteria) that are unable to carry through the act of genetic recombination has raised hopes of an early understanding of the biochemistry of crossing over. Two points of interest emerged early in this work: several enzymes are involved, and some regions of some chromosomes are recombined by enzymes especially adapted to act upon those regions. In a general sense the enzymes involved in recombination presumably include enzymes which cut DNA molecules, enzymes which remove stretches of one chain of a duplex in order to expose the base-pairing surface of the other (followed by joining by hydrogen bonding of two fragments whose complementary chains have been exposed through the same region), enzymes which replace missing single-chain stretches if necessary, and enzymes which complete the covalent union of the chains donated by the two parents. Enzymes potentially competent to carry out each of these steps have been identified, and it should not be much longer before we know the fashion in which these or similar enzymes work together to bring about crossing over.

Fine-Structure Analysis

Despite our uncertainties regarding its mechanism(s), recombination has been a valuable analytical tool. By means of an exhaustive analysis of the properties of *rII* mutants in T4, S. Benzer clarified our picture of the nature and behavior of genic material.

In T4, the *r* mutants are detected by the morphology of the plaques which they make when plated with strain B of *E. coli*. An appreciable fraction of the *r* mutants so detected fail to make plaques when plated on strains of *E. coli* which harbor the prophage λ (see Chap. 8). Strain K was the prototype of strains upon which some *r* mutants fail to grow; such mutants were called *rII* mutants. In crosses of *rII* mutants to other T4 mutants, Benzer determined that all of the *rII*'s fell within one rather small segment (corresponding to about 1 percent of the circular map of T4) raising the possibility that they were all identical reoccurrences of a particular mutation. However, by three criteria such was shown not to be the case. (1) The rates at which the various mutants reverted to wild-type (that is, to the ability to plate on K) differed. Reversion rates varied from as much as 10^{-3} or so to less than 10^{-8}. (2) Crosses conducted between independently isolated *rII* mutants (using strain B as host) usually resulted in the production of some wild-type recombinants. The wild-type frequencies observed were small (4 percent or less) confirming that the markers were located in the same region of the map. The observed wild-type frequencies obtained from a number of two-factor crosses were used to map the "*rII* region." The maps obtained were linear although the certainty of that conclusion was somewhat weakened by mapping difficulties from high negative interference. (3) Simultaneous infection of strain K by some pairs of independently isolated *rII* mutants led to phage production. The results of such pairwise "complementation" tests permitted the assignment of most of the *rII* mutants to either one or the other of two groups, called *A* and *B*. Members of the same group produced no phage when used together to infect K; when members from each group were used together they gave a good yield of phage when permitted to infect strain K simultaneously. The mapping data and the complementation data related to each other as follows: the group A mutants and the group B mutants were located in nonoverlapping but adjacent segments of the linkage map. The combined results of the mapping and the complementation data define (presumed) segments of the chromosome. Benzer called these segments *cistrons* in order to stress the operation that defined them. A proper complementation test involves the comparison of the yields from two different cultures infected with mixed types. In one case (the *trans* test) the culture is infected with several particles per cell of each of the two mutants in question. In the other case (the *cis* test), the culture is infected with several wild-type particles and simultaneously with several double-mutant particles. Particles which are mutant in the *same* cistron give more or less normal growth (that is, growth characteristic of wild-type phage) in the *cis* test while giving essentially no phage progeny in the *trans* test. Particles mutant in different cistrons, on the other hand, give about the same number of progeny particles in each of the two tests. This definition of a cistron corresponds pretty well with *one* of the meanings of the word *gene*. One can get overattached to either of these words (*cistron* because its rigorous operational definition has proved to be too rigid, and *gene* because its aristocratic though ambiguous past invokes loyalty to the word while beclouding the reality). We will opt for cistron here, but guardedly. In Chap. 10 we shall see that "cistron" can usually be equated with a stretch of nucleic acid responsible for the specification of the amino acid sequence of a particular polypeptide chain. And we'll look at the mechanism by which that specification is executed. But now, let's return to the *rII* region and its mutants for their usefulness in defining the structure and behavior of a relatively short stretch of genic material.

About 7 percent of spontaneously occurring *rII* mutations clearly involve alterations more extensive than a single base pair. The most conveniently determined property of high diagnostic value for these mutations is their apparent inability to undergo reverse mutation; their reverse mutation rate is certainly less than 10^{-10}. When these multisite mutants are employed in crosses, they give frequencies of recombinants indicative of their gross nature; they fail to produce wild-type recombinants with each of the members of a map sequence of markers all of which *do* recombine with each other. The presumptive molecular basis of these multisite mutants has been determined in several instances. The frequency of particles recombinant for loci r_a and r_b was measured in crosses of the following sort:

$$\text{I} \qquad\qquad \text{II}$$

$$\frac{r_a\ r_m^+\ r_b^+}{r_a^+\ r_m^+\ r_b} \quad \text{and} \quad \frac{r_a\ r_m\ r_b^+}{r_a^+\ r_m\ r_b}$$

where r_m is the multisite mutant and r_m^+ is its wild-type allele; loci r_a and r_b are each close to r_m. Crosses of type II give lower frequencies of recombinants for loci r_a and r_b, indicating that the multisite mutations are deletions. Measurements of the density of these multisite mutant particles, however, fail to reveal a decreased DNA content.

Streisinger and his associates have shown that the terminal redundancy in r_m-mutant particles is compensatorily longer than in r_m^+ particles.

Crosses involving only deletions in the rII region have provided an operationally unique demonstration of the one-dimensionality of a linkage map over relatively short distances. Two deletions that invade a common region of the map fail to give recombinants (by definition) when crossed with each other. In a one-dimensional map the relative positions of each of a number of deletions can be represented, for instance, like this:

```
              b                    e                    g        h
      _____       _____         _____   _____
  a                    d                     f
_____               _____               _____
           c                         j
        _____                   _____
                        i
                  _____
```

Such an array derives from the results of all the possible two-factor crosses shown here (+ = some wild-type recombinants produced; − = no wild-type recombinants produced).

		Other parent								
		a	b	c	d	e	f	g	h	i	j
One parent	a	−	−	−	+	+	+	+	+	+	+
	b		−	−	−	+	+	+	+	−	+
	c			−	−	−	+	+	+	−	+
	d				−	−	+	+	+	−	+
	e					−	−	+	+	−	−
	f						−	−	+	−	−
	g							−	+	−	+
	h								−	+	+
	i									−	−
	j										−

Is any other array compatible with these data?

Enthusiastic extension of this approach is notable for its failure to uncover a single matrix like that shown below left. Convince yourself that this significant failure tends to rule out topological map relationships of the sort shown below right.

```
        w   x   y   z
   w    —   —   +   —
   x        —   —   +
   y            —   —
   z                —
```

Thus, there is no need to involve two dimensions when constructing a map by this (or any other) procedure.

Organization of Immature ("Vegetative") Viral DNA

T4 is among the best studied and certainly among the most interesting with respect to the intracellular behavior of its chromosome. We shall document it as a case history with occasional references to the behavior of other phages.

The sequence of hundreds of marked loci on the map of T4 has been determined from examinations of recombinant frequencies arising in crosses. The map is linear, unbranched, and of finite length, but it has no ends! It is circular. The first crosses that clearly established this map configuration involved coinfection by particles marked at three loci (three-factor crosses). In phages, as in any other creatures, the map sequence of three linked loci is usually more convincingly established by an examination of recombinant frequencies from a single three-factor cross than by comparison of the recombinant frequencies from each of the three possible two-factor crosses involving the same loci.

When the sequence of three loci, say a, b, and h, was determined, it was found that they were linked in that order. When loci h, i, and q were marked in a three-factor cross, the order was again "alphabetical." (Please don't be confused by my after-the-fact alphabetical renaming of the loci that were actually employed. It will simplify the later parts of this discussion.) Crosses to determine the sequence of loci q, r, and y gave the result you've come to expect. When loci y, z, and a were marked, however, the sequence was clearly seen to be y–z–a, with locus a tightly linked to z. These results can be represented on only one kind of map, a closed curve, or circular one.

A physical basis for the linkage map circularity of T4 (and the other T phages of even number as well as the unrelated phage P22) is revealed by studies on the structure of the chromosomes of the mature phage particles. C. A. Thomas, Jr., and coworkers conducted some clever studies of T4 chromosomes among the best (for didactic purposes, at least) of which was the following. Heating of DNA duplexes (such as viral chromosomes) is known to disrupt the double helical structure and permit the unwinding and mutual separation of the two polynucleotide chains. Sufficiently slow cooling of the solution of disrupted duplexes permits the reformation of good double helical material. If the slow cooling (annealing) of the DNA is carried out on rather dilute suspensions, only bimolecular condensations are likely, and most of the single-chain material is converted back to duplexes of precisely the original structure. Since large numbers of duplexes are dealt with in such a reaction, a given single chain almost never anneals with the complementary chain with which it was originally associated. This is irrelevant as long as the original duplexes all had identical nucleotide-pair sequences (as does λ, for instance). In the case of T4, however, the melting and annealing of the *linear* DNA duplexes results in the production of *circular* duplexes (Fig. 7.4). Other experiments fully justify the suggestion given by this experiment that the nucleotide sequences of T4 chromosomes are circular permutations of each other. Some particles have their genetic loci arranged in the order a–z, others have sequences m–z–a–l, and so on.

In the yield from mixed infections by r and r^+ (wild-type), about 1 percent of the progeny particles are demonstrably "heterozygous" (that is, they give rise to both r and r^+ particles when they infect a bacterium). More systematic measurements of the heterozygote frequency have shown that for any r mutant differing from wild type by a minimal difference (presumably a single nucleotide pair), 1.4 percent of the particles from cells infected by equal numbers of r and r^+ phage are heterozygous. When the r mutant is a deletion (see Chap. 5 and below), only 0.4 percent of the offspring of a mixed infection are heterozygous. This value is independent of the extent of the deletion over a wide range. This result and others, coupled with a handful of preconceptions, suggest that there are two kinds of heterozygotes; "small" mutants can indulge in the formation of both kinds, but grosser mutants can participate in the formation of only one. In the early 1960s George Streisinger reasoned this and other observations into a promising hypothesis: only small mutants can form heteroduplex heterozygotes; presumably the degree to which the Watson–Crick base-pair requirements can be violated is limited. Deletion mutants, as well as small mutants, can form a second type of heterozygote. Streisinger's hypothesis explains the second type of heterozygotes by supposing that each chromosome in a circularly permuted population possesses a terminal redundancy of the initial loci. A chromosome whose basic sequence is a–z, for example, is more exactly written as a–za^+ if recom-

148

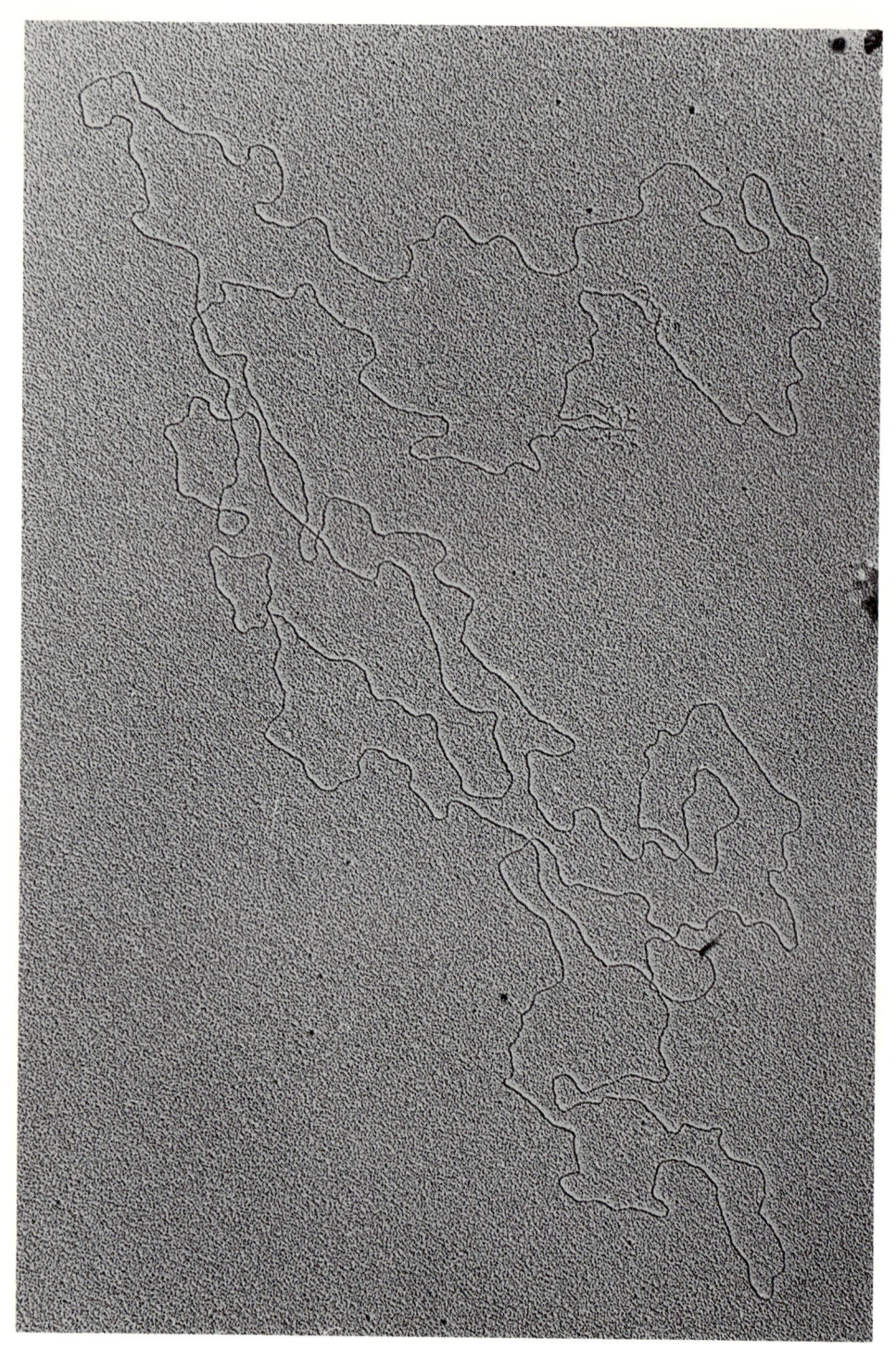

bination occurred between the terminal loci. The chromosome of such an individual would look like this:

$$\xrightarrow{}$$
$a \qquad\qquad\qquad za^+$
$$\xleftarrow{}$$

where the arrows, as usual, represent the two complementary chains of a DNA molecule. This idea provides an explanation for the origin from a single particle of the circularly permuted population. Streisinger's notion is that daughter chromosomes in an infected cell can synapse with each other like this:

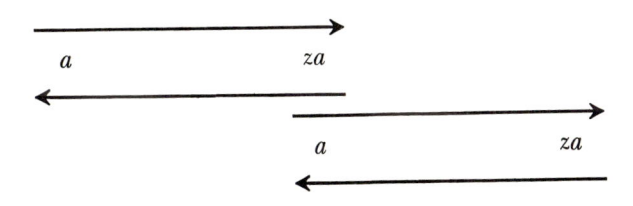

$$\xrightarrow{}$$
$a \qquad\qquad za$
$$\xleftarrow{}$$
$$\qquad\qquad\xrightarrow{}$$
$\qquad\qquad a \qquad\qquad za$
$$\qquad\qquad\xleftarrow{}$$

Exchange in the synapsed region can result in the formation of a "Siamese" chromosome. Reiterations of this process could lead to polymers of the polymer. Sometime before or during maturation, phage-sized chromosomes must be cut out of these monsters. The amount cut out would be a genome-plus-a-bit with starting points in the locus sequence well scrambled. The existence of terminal redundancy in mature T4 chromosomes has been clearly shown by physical methods. Furthermore, physical examinations of the intracellular state of T4 DNA have been compatible with this picture—the physical picture (Fig. 7.5) is as unclear as the one deduced from the genetic studies.

The smaller phage, λ, has been shown to occur in several distinct physical states within its host cell. The chromosome is injected from the mature particle as a linear molecule (not permuted) which is duplex throughout except for a stretch of about 20 nucleotides at each end. The two single-strand ends are complementary in sequence to

FIG. 7.4. *T4 DNA "circularized" by melting and annealing. When T4 chromosomes, which are linear, are heated in solution, the single chains of the duplexes separate. Slow cooling (annealing) permits the reformation of duplexes. In the case of T4, this procedure leads to the formation of circular duplexes because the original duplexes bear nucleotide sequences which are circular permutations of each other. The two little "bushes" near the upper right hand corner are single-chain DNA marking the redundant termini of the two single chains which united to form this duplex. The photo was kindly supplied by Lorne MacHattie and C. A. Thomas, Jr.*

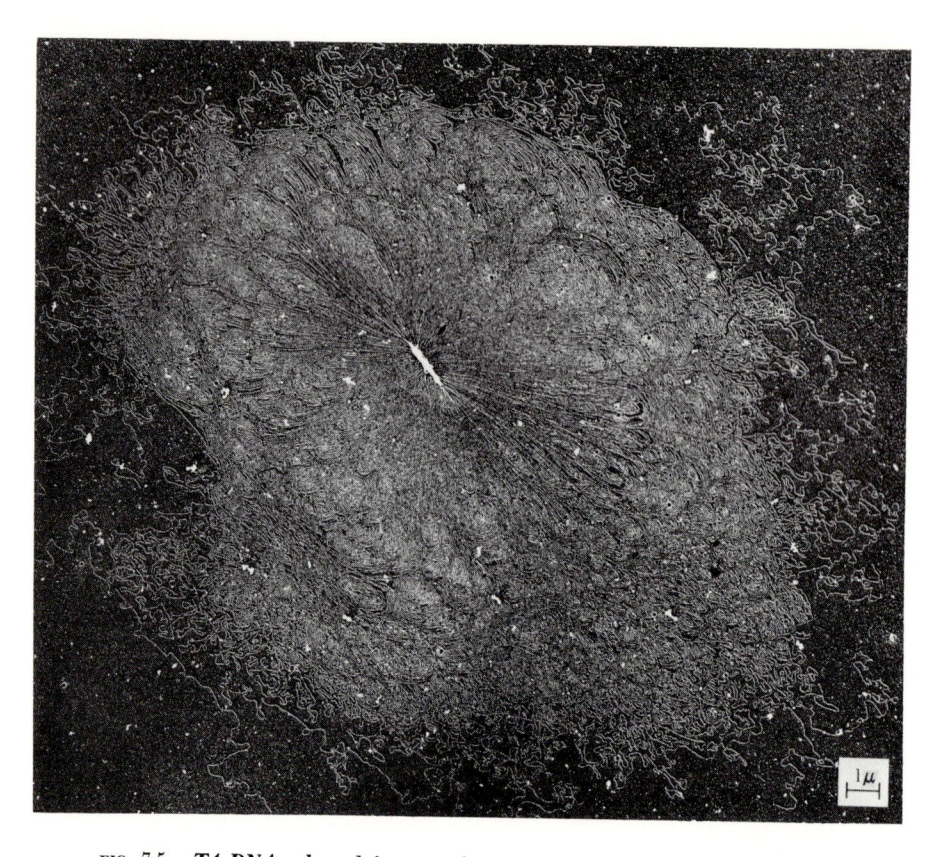

FIG. 7.5. *T4 DNA released from an infected cell by artificially induced lysis. The envelopes of infected bacterial cells were gently digested after T4 DNA duplication had proceeded for some time. High molecular weight DNA was partially purified by centrifugation and then mixed with the protein cytochrome c, which coats the DNA. The viscous solution containing protein-coated DNA was then run slowly down a glass slide onto an aqueous surface. The resulting DNA-cytochrome c film was picked up on an electron microscope grid and shadowed with metal. The doily in this beautiful picture by Joel Huberman represents the T4 DNA from a single infected bacterium. This picture was chosen over others for its beauty. It's only fair to point out that the T4 strain used in the infection is defective in its ability to make the major protein component of the phage head. Thus, the amount of free DNA in this infected cell is several times greater than in a wild-type infected cell. It is a characteristic of Dr. Huberman's pictures that the DNA appears to radiate from a "complex." By autoradiography the complex has been shown to be the major site of DNA synthesis.*

each other. By virtue of "spontaneous annealing" of these ends (followed by enzymatically catalyzed sealing of the two single-chain gaps), the linear molecule is converted into a circular one. Structures of molecular weight exceeding that of the mature λ chromosome arise also during the infection. Their structure and mode of origin are in doubt, but are not likely to remain so for long. Their relationship to the circular molecules is speculated upon, and you are invited to contribute, too. *One* role of the circular form of λ in the life cycle of that phage seems well established; it is an intermediate in the incorporation of the λ chromosome into the chromosome of the *E. coli* host cell. We'll look at *this* business in detail in Chap. 8.

Summary

The structures of virus chromosomes are well known relative to those of the chromosomes of larger creatures. This knowledge has provided the foundation for promising hypothesis regarding the mechanism of recombination between markers resident on the same chromosome. It has also provided the basis for our current picture of a "gene" as a linear segment of nucleic acid subject to mutation at each of a large number of sites which are separable from each other by genetic recombination. These rewards of the study of viral recombination were achieved in the face of apparently complex chromosomal life cycles.

References

Amati, P., and M. Meselson, "Localized Negative Interference in Bacteriophage λ," *Genetics, 51* (1965) 369–79. The description of clustered exchanges in phage λ by application of a general method for constructing a map whose distances are additive.

Benzer, S., "The Fine Structure of the Gene," *Sci. Amer., 206,* No. 1 (1962), 70–84.

Meselson, M., and J. J. Weigle, "Chromosome Breakage Accompanying Genetic Recombination in Bacteriophage," *Proc. Natl. Acad. Sci. U.S., 47* (1961), 851–68. Reprinted in *Papers on Bacterial Viruses,* 2nd ed., G. S. Stent, ed. (Boston: Little, Brown & Co., 1965), pp. 218–29.

Stahl, F. W., "Recombination in Bacteriophage T4, Heterozygosity and Circularity." *Symp. Biol. Hung., 6* (1965), 131–41. A review of the genetic observations leading to the picture of the terminally redundant, permuted chromosome of T4.

Problems

7.1. A student made two-factor crosses involving a number of pairs of markers in a newly discovered phage. She observed the following:

Loci marked in the cross	Observed frequency of recombinants (%)
AB	17.9
AC	24.9
AD	10.9
AE	22.9
AF	20.0
BC	24.5
BD	15.0
BE	17.9
BF	23.9
CD	25.0
CE	21.7
CF	21.7
DE	21.7
DF	21.7
EF	25.0

(a) Draw a linkage map for this phage showing the order of the six markers the student used. (For some of the loci mere inspection of the data permits you to assign correct relative distances.)

(b) Calculate the coefficients of coincidence for the following pairs of intervals: (1) AD and DB; (2) DB and BE; (3) AF and FC.

7.2. In phage λ, the observed recombinant frequency in crosses involving the two markers at opposite ends of the map is about 0.15. These markers are known to be very close to the opposite ends of the λ chromosome.

(a) Using the mapping function for λ in Fig. 7.2, determine the map distance (in Morgans) between the terminal markers.

(b) What frequency of recombinants is observed in a two-factor λ cross involving markers which are 1 percent of the total map length apart?

7.3. A student of Benzer's crossed a number of different deletion mutants in the *rII* region of T4. In each case, he infected cells of *E. coli* strain B with several particles of each of two different types and scored the progeny virus for presence or absence of wild-type recombinants. The matrix given here summarizes the results of these crosses. The parents involved in the crosses are indicated along the margins of the matrix. An entry of "+" in the matrix means that wild-type recombinants were produced, while "−" means that wild types were not produced.

Designation of the deletion in one parent

```
 1   2   3   4   5   6   7
 −   +   −   −   +   +   +   1
         −   +   −   −   −   +   2
             −   −   −   −   −   3   Designation of the deletion
                 −   −   −   −   4   in the other parent
                     −   −   −   5
                         −   +   6
                             −   7
```

A map derivable from the results of these crosses is shown below. Within the parentheses above each line, each of which symbolizes a deletion, put the number of the deletion which best fits.

Eight

Recombination in Bacteria

We have encountered genetic recombination among the bacteria in two previous instances. Bacterial transformation was described briefly for its importance in identifying the molecular basis of heredity. Bacterial conjugation provided the suggestion of a linear organization of chromosomes as revealed through experiments involving the interruption of chromosome transfer. In this chapter we shall examine these two processes further; we shall look also at several other mechanisms for juxtaposing DNA that result in recombination.

Phages and the Bacterial Chromosome

Not all strains of bacteriophages are fully destructive to their host cells. The temperate (as opposed to the virulent) strains of phages have two distinct modes of parasitic reproduction. Early in the course of infection of a bacterium a temperate phage chooses between two mutually exclusive paths of development. In some of the infected cells phage development proceeds lytically; a period of DNA duplication is followed by the appearance in the cell of mature phages with lysis occurring soon after. In others of the infected cells one (or sometimes more) of the phage chromosomes becomes added to the bacterial chromosome. Hand in hand with this act of lysogenization the host cells (and thus their descendents) acquire immunity to the lytic phase of phage reproduction; they survive not only that particular infection but also any subsequent infections by the same strain of phage. Different strains differ in the fraction of infections that lead to reduction to this prophage state (the state of the phage following lysogenization). The factors that influence the frequency of reductive responses are described in Jinks' *Extrachromosomal Inheritance,* another volume in this series. Here we are concerned with the physical nature of the interaction of the phage and bacterial chromosome; it bears on our understanding of DNA duplication, mutation, and recombination, and on the organization of DNA within chromosomes.

A prophage is duplicated along with the bacterial chromosome of its host, and the two daughter prophages are transmitted one into each daughter cell. The attachment of prophage to the bacterial chromosome is typically a secure one; only rarely in the normal course of events does the prophage abandon its liaison and enter the lytic (destructive) phase of its life cycle (see Problem 8.1). For the phage λ, and for several other well-studied strains of temperate phages, the locus of phage attachment is a unique one. In bacterial conjugation experiments the time of transfer of the prophage from an Hfr to an F^- cell is the same for each Hfr cell in a synchronously mating population of lysogenic (prophage-carrying) cells.

Transduction

When λ does leave the bacterial chromosome and become lytic, the particles that emerge from the host are generally indistinguishable from the particle that lysogenized the ancestor of the unfortunate cell. Rarely, however, particles emerge that differ from λ in two respects. (1) When infecting host cells by themselves, they are rarely able to enter the lysogenic phase and apparently never enter the lytic phase of reproduction (they are defective). (2) When they do become prophage, they may confer upon the cell a hereditary characteristic of the host from which the defective particles emerged. The only parts of the bacterial chromosome that λ can so transduce are the regions directing the synthesis of enzymes required for the fermentation of the monosaccharide galactose and for the production of the vitamin biotin. Reasonable hypotheses for the structures of the chromosomes of transducing particles and lysogenized cells have been arrived at from recombinational analysis.

The chromosomes of defective, galactose-transducing λ particles (λdg) can multiply normally in the presence of ordinary λ. A cell jointly infected with both kinds of particles may enter either the lysogenic or the lytic cycle of phage production. If the lysogenic mode is adopted, the cell may become lysogenic for both λ*dg* and λ. If it enters the lytic phase, it produces essentially equal numbers of the two kinds of particles. We shall examine the consequences of the lytic response first.

The map of λ*dg* can be determined from the results of crosses between lytic phase λ*dg* and λ particles marked at a number of loci. The map sequence of loci for λ*dg* is identical to that for λ except that a stretch of the map to the "left" of center is "missing"; the λ particles emerging from the cross carry each of the markers of the λ*dg* except for markers at those loci in the missing region. The defectiveness and the transducing capability of λ*dg* particles seem to have a simple common cause. It appears that λ*dg* has exchanged a stretch of its own chromosome for the galactose region of the chromosome of its host.

The portion of the λ map missing from λ*dg* particles varies. Whereas λ*dg* particles in a given clone are exactly alike, particles of independent origin may differ with respect to the length of the missing region of the map. It appears that one end of the missing region may be fixed while the other end is somewhat variable (see Fig. 8.1). This variability contributes to an observed difference in DNA/protein content (as inferred from density measurements) among mature particles of various λ*dg* strains. The correlation between map length of a missing segment and density is rather weak, however, suggesting that the length of the inserted piece of bacterial chromosome may be variable as well.

Our understanding of the organization of chromosomes would obvi-

FIG. 8.1. *A linkage map of* λ *indicating the region missing in* λdg *particles. The defective* λ *particles that can transduce the* Gal *region of the* coli *chromosome appear by genetic tests to be missing the region of the* λ *chromosome that corresponds to the part of the linkage map between* A *and* B *in the figure. The* A *end of the region has been demonstrated to vary from one* λdg *strain to another. The* B *end, on the other hand, has shown no sign of variability.*

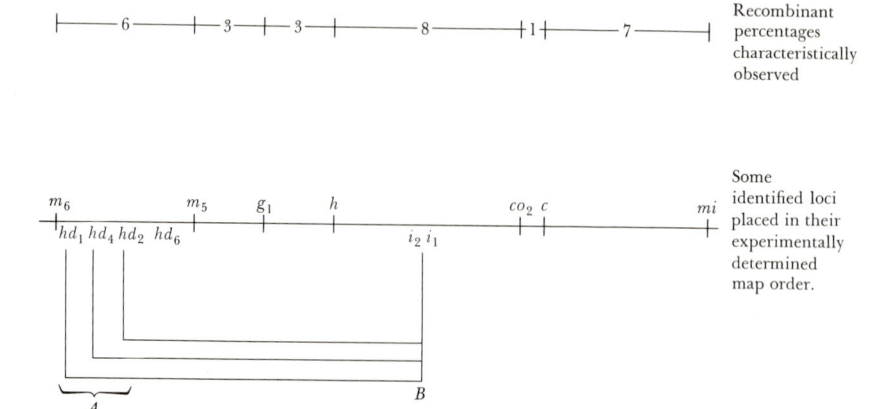

ously benefit from a deeper understanding of the nature of the association of prophage with bacterial chromosome and transducible markers with the phage chromosome. The latter relationship is relatively easy to examine; chromosomes isolated from λ*dg* appear by the applied physical criteria to be as simple as those isolated from λ. Thus, it seems likely that the transducible galactose region represents a physically ordinary stretch of the single DNA molecule that is the chromosome.

The physical relationship of prophage to the bacterial chromosome is less certain than is the relationship of transducing DNA to the chromosome of phage particles. At this time there appears to be no compelling reason to reject the structurally simple view that the prophage is inserted into the bacterial chromosome; from a physical point of view, the chromosome of a lysogenic cell differs from that of a corresponding nonlysogenic cell only by being a trifle longer.[1] A direct physical test of this simple view is made possible by the existence of bacterial strains which carry a small chromosome in addition to their normal chromosome. (We'll investigate the origin and nature of these extra little chromosomes later in this chapter.) In some strains the extra chromosome includes the region of attachment of λ, and λ can be added to them. In all respects, the lysogenized little chromosome has a simple (circular) Watson–Crick structure, and its molecular weight is greater than that of the nonlysogenized chromosome by exactly the molecular weight of the λ chromosome.

In 1960 E. Calef and G. Licciardello determined the sequence of three markers in the λ prophage. They conducted their "prophage crosses" by mating lysogenic male and female strains of *coli* and scoring the exconjugants for the markers carried by the prophage. Since most of the exconjugants carried either one or the other of the parental types of prophage, data on the relative frequencies of the possible recombinant types were meager. However, the data obtained argued that the order of the three markers on the prophage linkage map was *different* from the order of markers on the map of λ resulting from standard crosses. Other techniques for studying recombination between λ prophages have since amply supported that courageous conclusion. The prophage map in fact turns out to be a circular permutation of the order of the standard map, in agreement with the model for phage reduction proposed by A. Campbell. Campbell suggested that following injection of the λ chromosome into a host cell, the

---

[1] By several criteria λ does not seem to replace any parts of the host chromosome when lysogenizing. (1) The occasional cell strains that have lost their λ prophage are indistinguishable by all pertinent criteria from corresponding strains that have never been lysogenized. In particular, such "cured" strains are capable of being relysogenized, and the new lysogenic strain is indistinguishable from the original one. (2) No hereditary "defectiveness" is associated with lysogenization by λ. We shall see later that clearly contrary evidence exists for another phage quite unrelated to λ.

158 THE MECHANICS OF INHERITANCE

The vegatative λ chromosome...

circularizes...

and synapses with an homologous region (━) of the bacterial chromosome.

Crossing over...

results in...

the reduction of λ to the prophage state with a new sequence of loci.

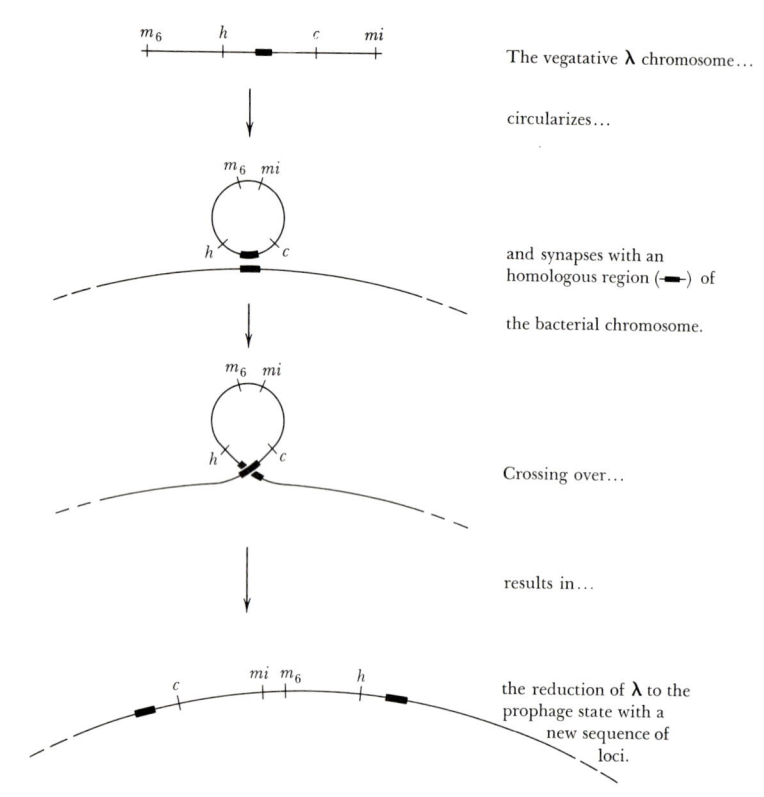

FIG. 8.2. *The mechanism for the reduction of λ to the prophage state as proposed by Allan Campbell in the early 1960s. Induction of λ is supposed to result from reversal of the reaction shown above. Double lysogens result from reiteration of the process with the result that the two prophages lie in tandem in the bacterial chromosome.*

chromosome can become circular. This step is now known to occur via annealing of the complementary single-chain ends of the mature λ chromosome. Then, Campbell proposed, the now circular chromosome undergoes a single crossing over with the bacterial chromosome. On the phage chromosome, the point of crossing over was taken to be some distance from the region of union of the ends of the mature chromosome; on the bacterial chromosome, the region of crossing over lies between the *Gal* and *Bio* regions. Campbell's model explained not only the difference between the prophage and the standard maps of λ, but also provided a simple explanation for the origin of particles able to transduce *Gal* or *Bio* markers of the bacterium. The model, diagrammed in Figs. 8.2 and 8.3, has become the cornerstone of our understanding of lysogeny.

Recombination in Bacteria

The reversibility of the incorporation of λ into the bacterial chromosome illustrates the possibility that many different bits of chromosome can cut in and out. While off the chromosome they may multiply extravagantly and even undergo recombination with their homologues.

FIG. 8.3. *A suggested mechanism for the formation of* λ *particles able to transduce the* Gal *region of the bacterial host (after a proposal by Allan Campbell in the early 1960s). According to this scheme,* λdg *formation is the result of a slipshod reversal of the reduction reaction (compare Fig. 8.2). The imprecision of the event is reflected in the variability from strain to strain of the end of the incorporated* Gal *region that lies close to* m_6 *(see Fig. 8.1). Particles which can transduce the* Bio *region of the* coli *chromosome (which specifies enzymes for producing biotin) presumably arise in a similar fashion.* Bio *lies just to the right of* λ *on the* coli *chromosome.*

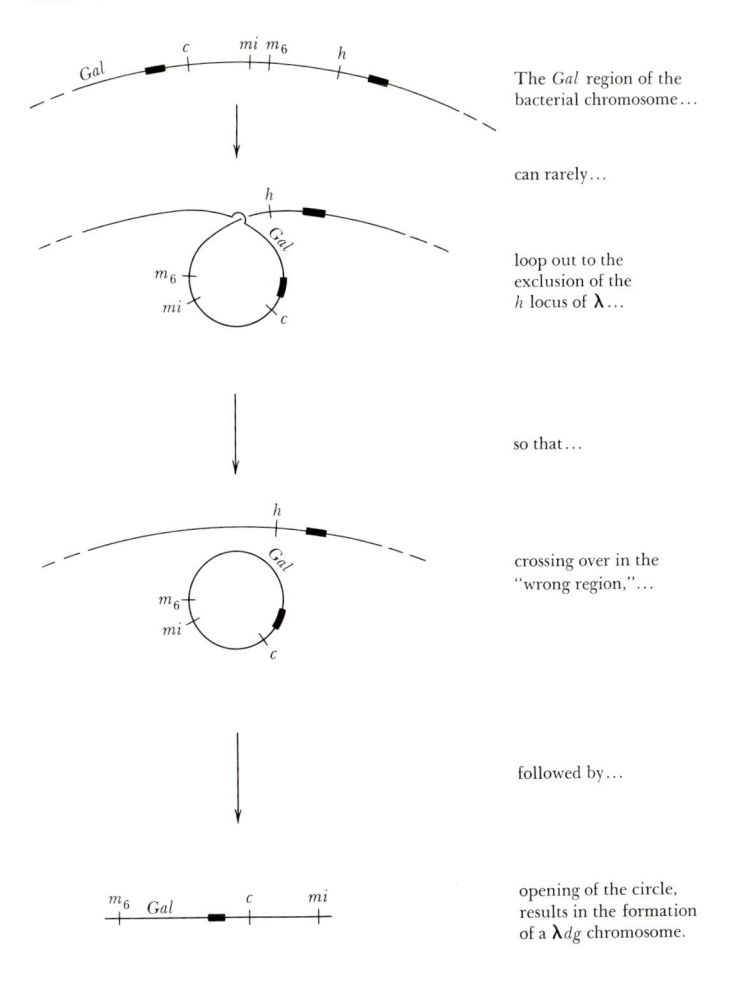

The Gal region of the bacterial chromosome...

can rarely...

loop out to the exclusion of the *h* locus of λ...

so that...

crossing over in the "wrong region,"...

followed by...

opening of the circle, results in the formation of a λ*dg* chromosome.

160 THE MECHANICS OF INHERITANCE

The significance of part-time chromosomal residents was emphasized by F. Jacob and E. L. Wollman (1961); they called such segments of genic material "episomes." Since episomes are (by definition) part-time inhabitants of the cytoplasm, they will come under discussion again in *Extrachromosomal Inheritance,* by Jinks, in this series.

There is a style of transduction recognizably different from transduction by λdg. Generalized transducers are temperate phages that can transduce any more-or-less short stretch of a host-cell chromosome. As in the case of λ, the transducing particles have been demonstrated to be defective for some phages. In the case of the phage P22, the transducing particles appear to carry exclusively bacterial DNA. Phage P22 is like T4 in that its chromosomes are circularly permuted, and its DNA exists in giant structures within the cell. As in T4, it seems likely that the determination of the size of the chromosome of the mature particle is determined by the amount of DNA which can be poured

FIG. 8.4. *A hypothesized crossover mechanism for reduction of the "mutator" phage mu-1. The two terminal regions of the phage chromosome synapse and cross over with homologous regions of the bacterial chromosome. A stretch of the bacterial chromosome is thereby replaced by the phage. Since the phage can become resident at any locus, there must be numerous points of homology along the bacterial chromosome. Perhaps these are nucleotide sequences that demarcate regions (cistrons) responsible for the specification of the amino acid sequence of distinct polypeptides.*

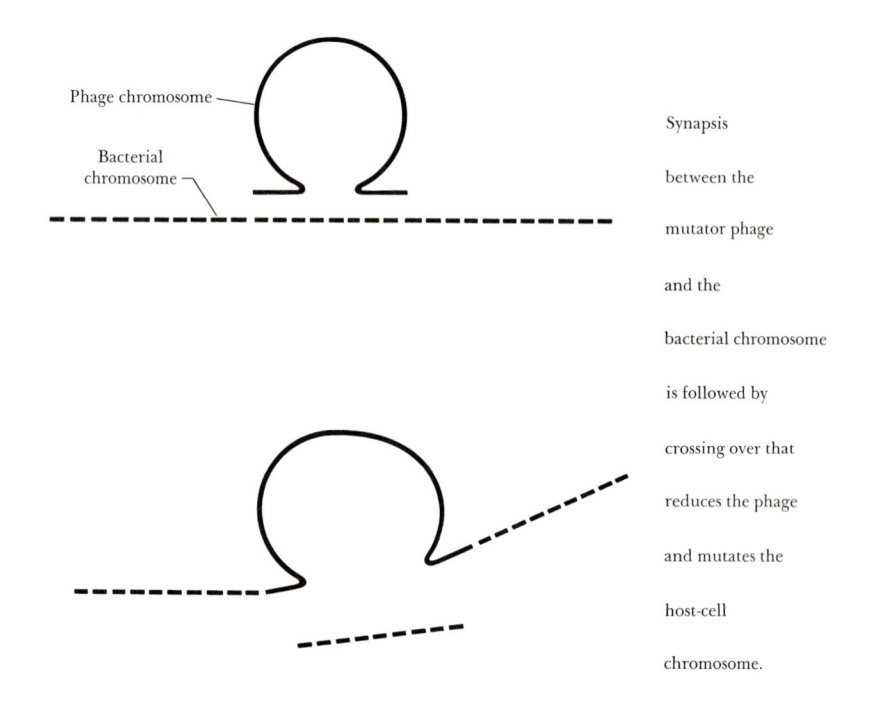

Synapsis between the mutator phage and the bacterial chromosome is followed by crossing over that reduces the phage and mutates the host-cell chromosome.

into the head. We would understand P22 transducing particles, then, as having had bacterial DNA poured into their heads "by mistake."

One phage that can induce hereditary alterations at (apparently) any locus of a bacterial chromosome has been discovered. These changes are not a consequence of transduction; the change induced is independent of the host upon which the phage was last grown. In every case the alteration induced is to a *loss* of a hereditary biochemical capacity. This mutation goes hand in hand with the reduction of the phage to prophage at the mutated locus. Two explanations seem reasonable. (1) The prophage prevents the proper operation of the region of the bacterial chromosome in its immediate vicinity. (2) The phage replaces a stretch of the bacterial chromosome upon becoming incorporated into it. A crossover mechanism like that diagrammed in Fig. 8.4 would have the observed consequences. How might one experimentally distinguish between these two hypotheses?

Episomes in Bacterial Fertility

In *Escherichia* and related genera the ability to conjugate and transfer genic material from cell to cell is dependent upon an episome, the F factor (fertility factor). Cells carrying the F factor (F+ cells) can pair with noncarriers (F− cells). During this union, a copy of the F factor is transferred to the F− cell, rendering that cell F+. What looks like a gratuitous exercise in sexuality is seen to be significant when we appreciate the ability of the F factor to interact with the *coli* chromosome. The F factor, a small circular DNA molecule, can enter the chromosome by crossing over in a manner analogous to the reduction of phage λ. Cells carrying F in the reduced state are truly fertile strains; during sexual union the migrating F factor carries a copy of the cell's chromosome along with it into the F− recipient. [These fertile cells are the *Hfr* (male) strains referred to in Chap. 5.] The integration of F into the *coli* chromosome is not perfectly stable. On occasion F reverts to its extrachromosomal state, and the ability of F to transport the *coli* chromosome is lost. Sometimes when F escapes, however, the presumptive crossover which eliminates the circular F from the chromosome occurs at a "wrong place" and the eliminated circle carries a stretch of bacterial chromosome along with the F DNA. Genetic markers carried on such modified F factors are rapidly transferred during sexual encounters between these F+ and F− cells. (The existence of modified F factors carrying the *Gal–Bio* region of the *coli* chromosome made possible the demonstration of the physical nature of the λ prophage state described earlier in this chapter.) We shall look at these modified F factors again in Chap. 11.

Jacob and Wollman studied a number of *Hfr* strains of *E. coli* K12 arising independently from a single F+ strain. Most of the strains isolated were unique with respect to the sequence in which they transferred markers during conjugation. However, the different sequences were related to each other by a simple rule: the sequences were circular permutations of each other with both clockwise and counterclockwise sequences represented. Only if conjugation were maintained until all the known markers were transferred did any of the F⁻ exconjugants become *Hfr*; that is, the F factor appeared to be the last marker transferred by each *Hfr* strain. These beautiful experiments led to the conclusion that F can be incorporated into the *coli* chromosome at any one of a large number of places and that the origin and direction of chromosome transfer is determined by the location and orientation of the reduced F factor.

Chromosomal transfer during conjugation requires chromosomal duplication; the nature of this requirement is currently the object of study primarily for the light that an understanding of it might shed upon the process of chromosome duplication per se. The way things stand now, it appears as if DNA synthesis in both the *Hfr* and the F⁻ is required at some stages to effect chromosome transfer. Furthermore, it appears that crossing over between the F⁻ chromosome and the originally transferred portion of the *Hfr* chromosome is required to permit completion of transfer. Briefly, the facts are these: (1) *Hfr* strains carrying incorporated F factors whose ability to duplicate is temperature-sensitive are unable to transfer chromosome at high temperature; (2) F⁻ strains whose DNA-synthesizing ability is temperature-sensitive are defective in their ability to *receive* chromosome from *Hfr* cells at high temperature; (3) transfer is depressed by the presence in the F⁻ cells of a large deletion located at the region of the chromosome corresponding to the originally transferred portion of the chromosome of the *Hfr* strain with which mating is attempted. The very tentative picture which emerges from these observations and conclusions (and some others) is the following: at least one of the chains of the circular male chromosome is cut at the F factor. The cut chain is displaced from the duplex by the synthesis of a new chain identical to it on the complementary chain. This synthesis proceeded stepwise from the point of the cut. However, the duplicating machinery is bound to the cell membrane near the conjugation tube so that one envisions the chromosome moving through the duplication machinery rather than the machinery moving along the chromosome. The displaced single chain is pushed (?) through the conjugation tube as it peels off the duplex. Once in the female cell, the single chain is converted to duplex DNA by the action of the female DNA-synthesizing enzymes. The resulting duplex undergoes crossing over near its tip with the circular female chromosome. The movement of the female chromosome through *its* duplicating machinery then pulls the single-chain released by the male through the conjugation tube. If you find this picture pretty hard to believe, I suggest you design some experiments which could test it.

The Mechanism of Recombination

Our knowledge of the structure of the products of recombination in bacteria is limited somewhat by the difficulty of performing structural studies on such large chromosomes. It seems sufficient to say that the known facts are adequately encompassed by overlap models like those we described for phage recombination in Chap. 7. Although this picture seems adequate for describing the recombination that accompanies bacterial conjugation, it may need modification in order to describe the recombination between transforming bacterial DNA fragments and the chromosome of the recipient cell. In some transformable species of bacteria, transformation can be accomplished by single-chain DNA derived from the donor's duplex DNA by heating. This and other observations suggest that transformation may often involve the incorporation of a single-chain stretch of DNA into one of the chains of the recipient chromosome with no alteration whatever in the complementary chain.

Recombination-Deficient Mutants

Mutants of bacteria which are unable to recombine were first described at about the time of appearance of the first edition of this book (1964). The description of the properties of these mutants has occupied those who hope to use the mutants to elucidate the chemical steps involved in recombination between DNA molecules. We should watch this work closely; it has already enlarged our picture of lysogenization. In phage λ, several classes of mutants with severely reduced ability to lysogenize have been studied. One of these mutants fails because it lacks part of the machinery required to undergo the incorporating crossover with the bacterial chromosome. Other mutants in λ *can* carry out that crossover with normal ability but have reduced ability to recombine markers in the rest of their chromosome. Mutants of λ defective in both of these recombination systems still recombine somewhat by virtue of the action of the host-cell recombination system. The study of the molecular basis of recombination is a challenging one, which needs students of physical chemistry, biochemistry, genetics —preferably those having interests and abilities in all three disciplines.

We introduced our discussions on recombination by an examination of the process as it occurs in meiosis. The chromosomes and their microscopically observable behavior provided a simple visual *conceptual* framework within which to think about random assortment and crossing over. We looked then at recombination processes in simpler creatures and found them to appear more complex! In Chap. 9, we look more closely at recombination of linked markers in higher organisms in an effort to describe the events in these forms at the level of

DNA duplexes. We will find there no shortage of complexities (apparent, at least), all of which are aggravated by our uncertainties regarding Chromosome structure.

Summary

The study of recombination in bacteria has elucidated the life cycles of not only the bacteria but also of a variety of little beasts (episomes) that cavort within them. The study of mutant bacteria (and episomes) defective in recombination will reveal the molecular mechanisms underlying the recombination of linked genetic markers.

References

Calef, E., and G. Licciardello, "Recombination Experiments on Prophage-host Relationships," *Virology, 12* (1960), 81–103.

Campbell, Allan, "Episomes," *Adv. Genet., 11* (1962), 101–45. Our knowledge of DNA interactions leading to recombination in bacteria is reviewed within the framework of the episome concept.

Clark, A. J., "The Beginning of a Genetic Analysis of Recombination Proficiency," *J. Cell. Physiol., 70,* Suppl. 1 (1967), 165–80. A review of studies on bacterial mutants with impaired ability to undergo genetic recombination.

Cuzin, François, Gerard Buttin, and François Jacob, "On the Mechanism of Genetic Transfer During Conjugation of *Escherichia coli,*" *J. Cell. Physiol., 70,* Suppl. 1 (1967), 77–88.

Jacob, François, and E. L. Wollman, *Sexuality and the Genetics of Bacteria.* New York: Academic Press, Inc., 1961. This book discusses in detail bacterial conjugation and lysogenization as illustrated by the outstanding work of two microbial geneticists.

Weigle, J., M. Meselson, and K. Paigen, "Modified Density of Transducing Phage Lambda," in *Structure and Function of Genetic Elements,* Brookhaven Symposia in Biology: No. 12 (1959), pp. 125–33.

Problems

8.1. A bacterial strain can be called lysogenic only if occasional cells release infectious phage particles. (This statement implies a corollary; many, perhaps most [or all] bacterial strains carry prophage-like chromosomal symbionts that remain undetected because of their failure to escape from the host-cell chromosome and be released as infectious virus.) The rate of release of phages by lysogenic cells can be measured. This problem illustrates one method of determining this rate.

Recombination in Bacteria

(a) Suppose that a lysogenic culture is adjusted to a concentration of 100 cells per ml; 1-ml volumes are then delivered into each of a large number of tubes. Each of these cultures is then permitted to grow to a concentration of 10^6 cells per ml. At that point each tube is examined for the presence of mature, infectious phage particles: 87 percent of the tubes are found to contain some extracellular phage particles. Calculate the probability of phage release per cell per duplication.

(b) The rate of phage release by the cells of a lysogenic culture can be altered by environmental agents. Ultraviolet light (UV) is a convenient agent for inducing lysogenic cells to liberate phages. A cell induced by UV lyses as a consequence of phage production. UV is lethal to bacterial cells for other reasons; the proper functioning of DNA is generally sensitive to UV. For the hypothetical results presented below calculate the fraction of cell-lethal hits that are a result of phage induction. Your answer, I trust, confirms the typical conclusion that UV killing of lysogenic cells by induction is far greater than all the other sources of UV killing combined.

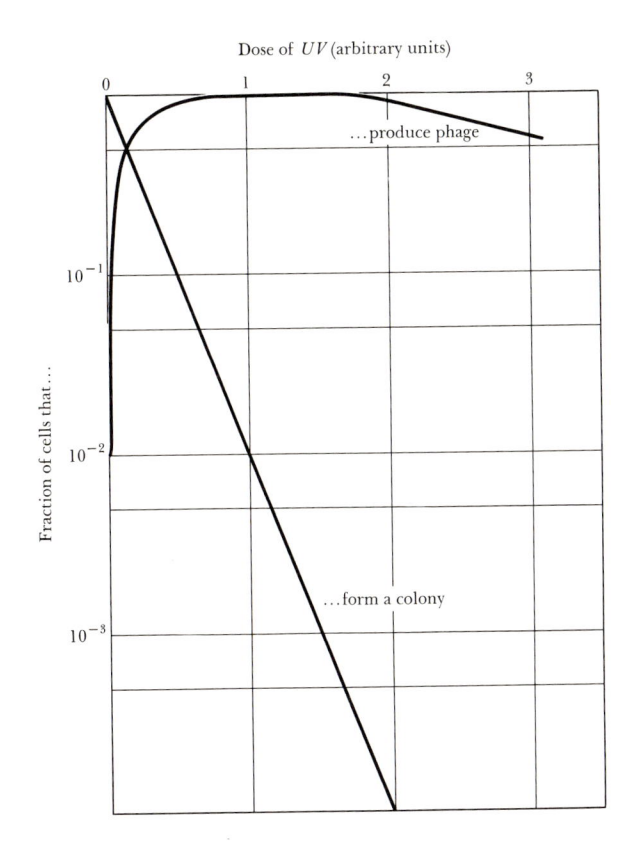

166 THE MECHANICS OF INHERITANCE

8.2. It is sometimes difficult to sustain conjugation in *E. coli* long enough for the entire genome to be transmitted from donor to recipient. Thus, in some crosses of an *Hfr* strain by an F⁻ strain, the linkage relations of only a fraction of the genetic markers distinguishing the two strains may be determined. Suppose that in such crosses the following *Hfr* strains donate the markers shown in the order written. (The order can be determined by interrupting conjugation with an electric blender.)

Hfr strain	Markers donated in order
1	Q W D M T
2	A X P T M
3	B N C A X
4	B Q W D M

Draw the presumptive sequence of markers in the F⁺ strain from which these *Hfr* strains were derived.

8.3. A student has a strain (strain 2) of bacteria that is resistant to drugs *A, B, C,* and *D*. He also has the sensitive strain (strain 1) from which the resistant strain was derived by four independent mutational steps. He can extract DNA from strain 2 and apply it to strain 1. Then he can plate the treated cells of strain 1 on agar containing various combinations of the drugs. Suppose he does that with the following results:

Drug added to agar	Number of colonies	Drug added to agar	Number of colonies
None	10,000	BC	51
A	1,156	BD	49
B	1,148	CD	786
C	1,161	ABC	30
D	1,139	ABD	42
AB	46	ACD	630
AC	640	BCD	36
AD	942	ABCD	30

He concludes that three of the markers conferring drug resistance are so close together on the bacterial chromosome that they are often located on a single DNA fragment. In transformation jargon, these are said to be linked.

(a) Which marker is *not* linked to the other three?
(b) In what order do the three linked markers occur?

8.4. The density of protein in CsCl is about 1.3. For DNA, the density in CsCl is 1.7. The density of phage λ in CsCl is 1.5. Assume that λ is composed solely of DNA and protein (with no air bubbles).
(a) Calculate the fraction of the particle volume which is DNA.
(b) Calculate the fraction of the particle mass which is DNA.
(c) J. Weigle, M. Meselson, and K. Paigen (in 1959; see References) observed a strain of λ*dg* which was less dense than ordinary λ by 0.0132 density units. What was the DNA content of these transducing particles relative to the DNA content of ordinary λ? (This one looks easy, but may be harder than it looks.)

Nine

Recombination in Higher Organisms: II

In Chap. 6 we described recombination of linked (as well as unlinked) markers as it occurs within the orderly meiotic process. Our analysis was essentially formal; the observations which we described gave us no hints as to the molecular mechanisms underlying an act of crossing over. In Chaps. 7 and 8 we investigated recombination in viruses and bacteria. In these cases, we were often unable to give a clear description of the rules governing the recombination process. However, we did emerge with at least a rough picture of the molecular structure of a recombinant chromosome and some notions regarding the steps (at the molecular level) involved in its formation. In this chapter we shall return to the study of recombination in higher organisms. We shall find that when we look closely we can see phenomena reminiscent of some of those observed for the viruses. Encouraged by the parallels, we shall speculate about the *molecular* basis of crossing over in higher organisms. Much of the work which we shall discuss has been conducted with the ascomycete *Neurospora crassa*. Our subsequent discussion will be easier to follow if we review the essential features of the *Neurospora* life cycle at this point.

Ascospore Formation in Neurospora

Neurospora proliferates in the haploid state. As in *Chlamydomonas,* the diploid state is short-lived. Nuclear fusion is followed promptly (or is immediately preceded by?) the premeiotic round of DNA duplication. The two successive meiotic divisions proceed in a long sack (ascus) sufficiently narrow that the resultant haploid nuclei stay in place. Thus the two nuclei at one end of the ascus are sisters derived from the same second meiotic division. (In Problem 6.6 we exploited this feature of the *Neurospora* life cycle to explore the problem of sister-strand crossing over.) In *N. crassa*, each haploid nucleus then undergoes a postmeiotic mitosis; this nuclear division is likewise conducted without queue-jumping, so that each original haploid product of meiosis is now represented at the appropriate place in the ascus by two usually identical haploid products of the postmeiotic mitosis. Each of these haploid nuclei then becomes encased in a spore wall. (It may be relevant to some queries that these ascospores are binucleate as a result of a mitosis occurring shortly after the postmeiotic mitosis.) These ascospores can then be examined in either of two ways. The more convenient way is to collect spores liberated by the plant from its mature, bursting asci. A fraction of this spore sample can then be analyzed for the frequencies of various types emerging from a cross. Different, and in some ways more powerful, information can be obtained by isolating asci before they burst, tearing them open (without disturbing the order of the spores in the ascus) and scoring the eight spores as to type. We'll start with a random spore analysis which can be used to illustrate high negative interference in *Neurospora*.

High Negative Interference

In Chap. 6, in our previous discussions of recombination of linked markers in higher organisms, we pointed out that coefficients of coincidence less than unity are typically observed. This interference is generally ascribed to "mechanical" properties of the Chromosomes. It stands in contrast to the results we observed in λ; for λ the coefficients of coincidence were all greater than unity. However, just as with λ, coefficients of coincidence in higher organisms increase sharply with decreasing distance at very small distances.

Consider the four-factor cross $Am_1 + B \times a + m_2 b$ where m_1 and m_2 are very tightly linked (giving, say, < 0.1 percent recombinants) and are in the same or neighboring cistrons, while A and B are roughly equidistant from m_1 and m_2 and give several percent (say 2 percent to 20 percent) recombinants with each other. In crosses of this sort, markers m_1 and m_2 generally represent mutations to a nutritional deficiency; thus, even though recombination between them is rare, large numbers of wild-type (that is, $+ +$) recombinants can be selected by their ability to grow on a medium lacking the relevant nutrient. The medium used is designed such that the four possible types of spores with respect to A, B, a and b can all grow. We may state the finding of high negative interference by saying that among the + + recombinants, the four types with respect to A and B often occur with appreciable frequencies. We'll look more closely at some of these frequencies later; for now we note merely that many of the + + recombinants carry A, others carry b, and some are $A + + b$ recombinants. By having selected for recombination between the two close markers m_1 and m_2, we seems to find ourselves in the midst of a cluster of exchanges. We can convince ourselves that the high coefficients of coincidence observed in these crosses are characteristic only of recombination in a very short region. We can propagate a sample of spores from the same cross on a medium supplemented with the nutrient for which m_1 and m_2 are auxotrophic. Then we can score the four possible types with respect to A and B. The result is not unexpected: less than half of the spores are recombinant for A and B, and the coefficient of coincidence for the production of the double recombinants $am_1 + b$ and $A + m_2B$ is less than unity. Thus, genetic exchanges in *Neurospora* come in "clusters" as they do in phages.

We discussed evidence that related genetic exchange in phages to "overlap" regions involved in the union of segments of chromosome derived from two different parental chromosomes. Our knowledge of phage chromosome structure made possible this connection between an observed genetic result and a defined molecular model. As we discussed in Chap. 5, our knowledge of Chromosome structure is minimal. However, we did find the simplest model for a Chromosome useful in explaining the semiconservative nature of Chromosome duplication. We shall keep to simple models in thinking about the recombination of linked markers, and will find that by doing so the picture of recombination which emerges shares with that for viral recombination not only the feature of high negative interference but also the involvement of overlap regions in recombinant formation. Our picture depends heavily upon the analysis of individual asci in four-factor crosses of the sort we just used for random-spore analysis.

Aberrant Segregation

In most of the asci from a *Neurospora* cross each allele is present in both members of each of two pairs of ascospores. (This is the 2 : 2 segregation of alleles which we discussed in Chap. 6 for *Chlamydomonas*.) Rarely, however, asci are found in which the 2 : 2 rule is violated. These violations may take any one of five forms; they are illustrated (and named) in Table 9.1 along with a 2 : 2 ascus. (The grouping of the spores in pairs is relevant to our present discussion, but the order of the pairs is not.) In crosses of two very closely linked markers, it is often the case that asci which contain recombinant

TABLE 9.1 *Types of Segregation in the Cross* $m \times +$

	Aberrant				Normal
6 : 2	5 : 3	3 : 5	2 : 6	4 : 4 odd	
+	+	+	+	+	+
+	+	+	+	+	+
+	+	+	m	m	+
+	+	m	m	+	+
+	+	m	m	+	m
+	m	m	m	m	m
m	m	m	m	m	m
m	m	m	m	m	m

spores manifest aberrant segregation for one or the other (rarely both) of the markers involved. Thus, for the case of aberrant segregation for marker m_1, the most frequently observed types of asci in which recombinant spores are found in the cross $+m_2 \times m_1+$ are shown in Table 9.2. (Again, the order of the spore pairs is not relevant to our

TABLE 9.2. *Types of Asci with Spores Recombinant for Two Close Markers*

Aberrant segregation for marker m_1 (but not for marker m_2)					Normal tetratype ascus
$+m_2$	$+m_2$	$+m_2$	$+m_2$	$+m_2$	$+m_2$
$+m_2$	$+m_2$	$+m_2$	$+m_2$	$+m_2$	$+m_2$
$+m_2$	$+m_2$	$+m_2$	m_1m_2	m_1m_2	+ +
$+m_2$	$+m_2$	m_1m_2	m_1m_2	$+m_2$	+ +
+ +	+ +	m_1+	m_1+	+ +	m_1m_2
+ +	m_1+	m_1+	m_1+	m_1+	m_1m_2
m_1+	m_1+	m_1+	m_1+	m_1+	m_1+
m_1+	m_1+	m_1+	m_1+	m_1+	m_1+

present discussion.) Let us examine the six kinds of asci above for common and distinctive features and their implications.

In *all* of the asci, two of the four spore pairs are unmixed and of opposite parental type. This observation is harmonious with our conclusion of Chap. 6 that only two of the four chromatids in a bivalent engage in crossing over in any particular short region. The 6 : 2, the 2 : 6, and the normal asci contain no mixed spore pairs; the 5 : 3 and the 3 : 5 as well as the 4 : 4-odd contain mixed spore pairs. The existence of mixed spore pairs (postmeiotic segregation) implies the involvement of heteroduplex structures.

In all the aberrant asci except for the 4 : 4-odd, only one of the two recombinants for loci m_1 and m_2 are present. Thus, recombination between very close markers is frequently a nonreciprocal process.

In phages, we found that heteroduplex heterozygotes were often recombinant for markers which bracketed the heteroduplex region. The random spore analysis combined with the ascus analysis we have made in our four-factor *Neurospora* cross leads us to suspect a similar relationship in higher organisms. If we examine the behavior of A and B in the aberrant asci our suspicion will be confirmed. Among aberrant asci, A and B are frequently recombinant (in a classical tetratype fashion). The spore pairs which are recombinant for A and B usually include the pair which is nonparental for m_1 and m_2; the frequency of recombinants among these nonparental pairs is about 50 percent. The recombinant frequency for A and B, though high in the aberrant asci, is even higher in those asci from the same cross which are reciprocally recombinant for m_1 and m_2. In asci which are classically tetratype for m_1 and m_2 it has been shown (most cogently by S. Fogel and D. D. Hurst in yeast) that A and B are almost invariably recombinant, with the + + spore pair being aB and the m_1m_2 spore pair being Ab. In other words, those acts of recombination between m_1 and m_2 which give normal segregation give the classical expectation of essentially no additional recombination in short adjacent regions.

Asymmetric Recombination

Noreen Murray (1968) determined the frequencies of the four types of outside marker combinations among the + + recombinants. She found that the frequency of recombinant types for A and B were about equal to the frequency of parental types. As expected, she found the aB recombinant to be more frequent than the Ab one. Unexpectedly, however, she found that the two parental classes (AB and ab) were unequal. For the two close markers m_1 and m_2 she used different markers in the same cistron. She found that the magnitude of the inequalities in these parental and the recombinant classes were different for different close marker pairs. Furthermore, those marker pairs which gave the most unequal parental ratios gave the most unequal recombinant ratios as well. The magnitudes of the inequalities appeared to depend on the location of the markers m_1 and m_2. Of the set of markers studied, those mapping near one end of the set gave larger inequalities than those mapping near the other end of the set. Let us examine a purely geometric interpretation of these data before we look at related data from ascus analysis.

172   THE MECHANICS OF INHERITANCE

The asymmetric recombination data can be "explained" in formal terms by supposing that synapsis of chromatids occurs only along short, structurally predetermined segments. (The number of such segments must be large to account for the large number of places at which recombination can occur on a Chromosome.) A few (say two or three on the average) occur between the two chromatids within a synapsed segment. The multiple exchanges, of course, are the source of the high negative interference; the asymmetry in the two classes which are parental for outside markers occurs whenever m_1 and m_2 are asymmetrically disposed within the fixed pairing region. The magnitude of asymmetry in the recombinant classes will vary with the position of m_1 and m_2 also. For instance, the configuration shown here (Diagram I) would lead to maximal asymmetries in the parental and the recombinant classes; among the $++$ recombinants, aB would

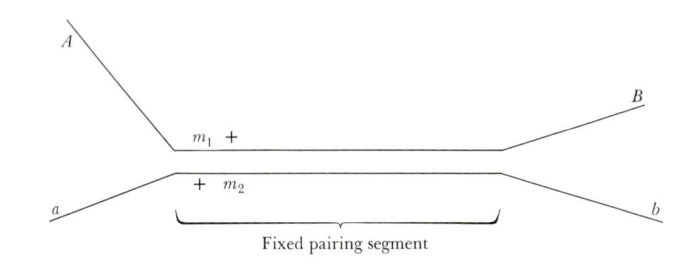

DIAG. I   Fixed pairing segment

exceed Ab and ab would exceed AB. As the marker-pair m_1m_2 is moved to the right, the asymmetries become less pronounced. For the case of the parental types, the asymmetries will reverse when the m_1m_2 marker pair is to the right of center of the fixed pairing segment.

Asymmetric recombination is manifested in asci as an asymmetry in the frequency of aberrant segregation. In a cross $m_1+ \times +m_2$ a majority of the asci bearing $++$ spores are typically the result of aberrant segregation at *one* of the two loci.

Now let us look at the results from five-factor crosses. After that we shall try to build a model.

Consider the cross $Am_1+BC \times a+m_2bc$ which is like our four-factor cross except that we have added the marker C to the right of B. D. R. Stadler noted the following properties of such a cross in a random spore analysis in *Neurospora:* in the overall population, recombination between A and B interfered with recombination between B and C (that is, the coefficient of coincidence was less than unity). Among the $++$ recombinants the same interference was observed for simultaneous recombination of A and B and of B and C. The complement of this observation was also true among $++$ recombinants which were parental types AB or ab, recombination between B and C was *not* depressed. In terms of fixed pairing segments, this result says that an odd number of exchanges in the fixed pairing segment results in positive interference in the region outside while an even number of exchanges has no such effect. Ascus analysis in yeast by S. Fogel and D. D. Hurst provided further information on these interference relationships. These authors showed that the exchanges between *B* and *C* manifested no chromatid interference. Among the + + spores which were *AB* or *ab*, exchanges in the *BC* region could involve any two (nonsister) chromatids with equal probability. Now let us think about a model. The model that we will build may not account for all the relevant published observations; furthermore, some of its features are somewhat arbitrary. However, the building of it will focus our attention on the primary observations and their relationships to each other, and the finished product can stand as a summary of this chapter. If you do not like my model, please try to build one of your own!

Before we start building, we list the observations that we hope to explain. We can relate them to the cross $Am_1+BC \times a+m_2bc$.

(1) Recombination of *A* and *B* is usually reciprocal and occurs between two of the four nonsister chromatids in a bivalent.

(2) Recombination of m_1 and m_2 is often nonreciprocal and usually occurs between the same two chromatids which are involved in the *A–B* recombination.

(3) Nonreciprocal recombination is typically manifested as 6 : 2 or 2 : 6 segregation of one or the other marker. The occurrence of 5 : 3 and 3 : 5 asci suggest the occasional involvement of heteroduplex overlaps (such as those we discussed in Chap. 7).

(4) Among + + recombinant spores the frequency of parental types (*AB* and *ab*) is about equal to the frequency of recombinant types (*aB* and *Ab*).

(5) Among + + recombinants, the frequencies of *ab* and *AB* are unequal by magnitudes which correlate with the inequalities of the frequencies of *aB* and *Ab*.

(6) In those asci in which recombination of m_1 and m_2 is reciprocal, *A* and *B* are almost always recombinant. When m_1 and m_2 give classical tetratype tetrads, high negative interference is not observed.

(7) Among + +, the recombination of *A* and *B* results in positive interference for further exchanges between *B* and *C*. This positive interference is lacking among the + + spores which are parental types *AB* or *ab*.

The Model

Suppose that the process of synapsis stimulates (or is dependent upon) DNA duplication (à la Watson–Crick, to keep things simple) at a number of spots along the Chromosome. We are interested in the occasions on which this duplication is stimulated in the region of m_1 and m_2. (The possibility of synapsis-stimulated duplication is suggested to me by the sex-stimulated duplication of the *E. coli* chromosome which we discussed in Chap. 8.) Then, our synapsed bivalent looks like Diagram II. (The scale is very distorted; please focus on the

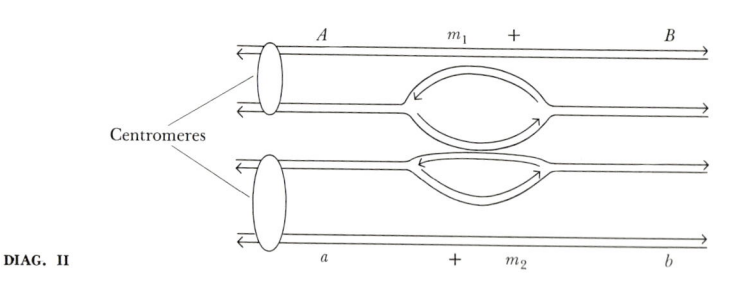

DIAG. II

topology.) The two circles now exchange parts with each other.

We suppose that a single exchange involves a breakage and reunion of two homologous arcs. Of the two conceivable reciprocal products of this process, only one can be used. This product has a structure like that of a phage recombinant. The event may be depicted as in Diagram III. Presumably the events implied by the arrow involve steps like

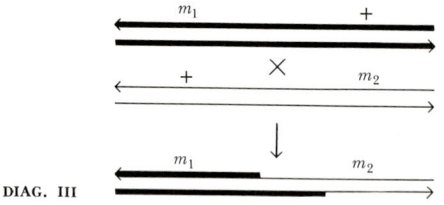

DIAG. III

those discussed in Chap. 7—DNA is cut, stretches of one chain of a duplex are removed to expose the pairing surface of the other, complementary stretches from two different molecules are annealed, and missing single-chain stretches are filled in. As shown in Fig. 9.1, this particular sequence of steps permits the formation of only one of the two conceivable recombinant molecules. In standard phage crosses such seems to be the case. The yields from individual infected cells contain reciprocal recombinants in uncorrelated numbers. Within the framework of our model, the feature of nonreciprocality of the elementary exchange event is closely related to the assumption of the sex-stimulated local duplication; if recombination between duplexes is nonreciprocal, the local duplication is essential to avoid the loss of entire chromatids.[1]

---

[1] Phage λ appears to undergo *reciprocal* recombination in two situations. The crossover which integrates λ into the bacterial chromosome must be a reciprocal one. Occasionally prophages carried on

If two DNA duplexes are cut (enzymatically, by an "endonuclease") at different nearby points...

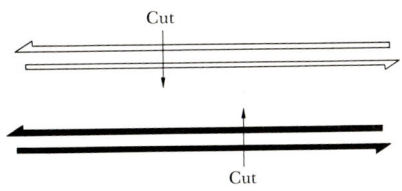

...and then subjected to single-chain nibbling (by an "exonuclease") in one chemical direction,...

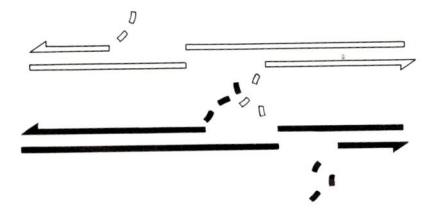

two of the four resulting pieces can rejoin by complementary base pairing. The other two pieces have no sequences complementary to each other, and are therefore "lost."

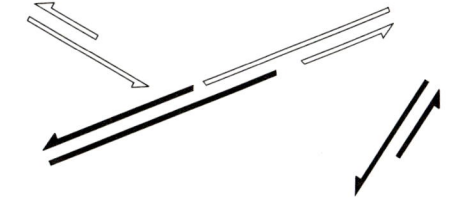

FIG. 9.1. *A plausible sequence of steps in recombination between DNA duplexes, which is physically nonreciprocal.*

The nonreciprocal exchanges involve the formation of overlap regions as in phages. When an overlap falls upon a marker, a postmeiotic segregation (a mixed spore pair) results. Thus arise the tetrads containing + + spores which manifest 5 : 3 segregation for one of the markers. We may note that this fraction should increase with decreasing dis-

---

a chromosome and an F episome respectively recombine with each other during multiplication of the host-cell population. M. S. Meselson (1967) showed that reciprocal products are frequently found together in single cells descended from the rare cells in which such an exchange occurred. These two observations have been taken as evidence against the nonreciprocal nature of exchange between DNA duplexes. For the sake of stirring things up, we'll take those observations to mean that in those two circumstances λ undergoes at least a local duplication, and that the ensuing events are in fact similar to those which we hypothesize to account for recombination of linked markers in higher organisms.

176   THE MECHANICS OF INHERITANCE

tance between m_1 and m_2. The remaining properties of the model emerge (or are imposed) without further reference to DNA duplexes. We can therefore simplify our diagram (IV). The circles correspond

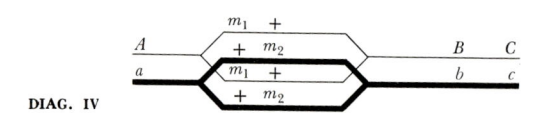

DIAG. IV

to fixed pairing segments, so that we may expect the asymmetries which reveal fixed pairing segments to emerge from the model. Two features of the model demand that each engagement of the circles involves *two* exchanges. (1) Circles need even numbers to avoid tangles (as we discussed in Chap. 7). (2) Two nonreciprocal exchanges will reduce the amount of DNA back to the duplex level in each chromatid. The various consequences of the two exchanges are shown in Diagram V. Let

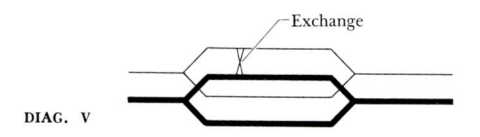

DIAG. V

us ignore the possibility of exchange between sister arcs. Then let the first exchange occur between any pair of nonsister arcs. Only one product (either one) survives; the other is consumed (Diagram VI).

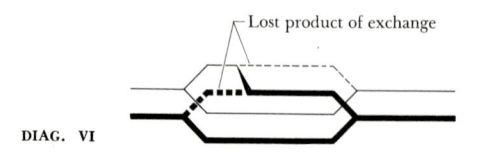

DIAG. VI

The second exchange, which we have arbitrarily placed to the right of the first, now occurs between either pair of nonsister arcs (Diagram VII). If the two possibilities are equally probable, then with equal

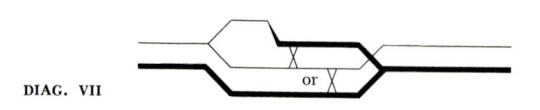

DIAG. VII

probability we retain the outside markers in parental combinations (VIII) or recombine them (IX). When we remove the kinks from the drawings, we see that we have pairs of chromatids which could manifest nonreciprocal recombination for close markers while giving reciprocal products for less tightly linked markers (X). Things came out

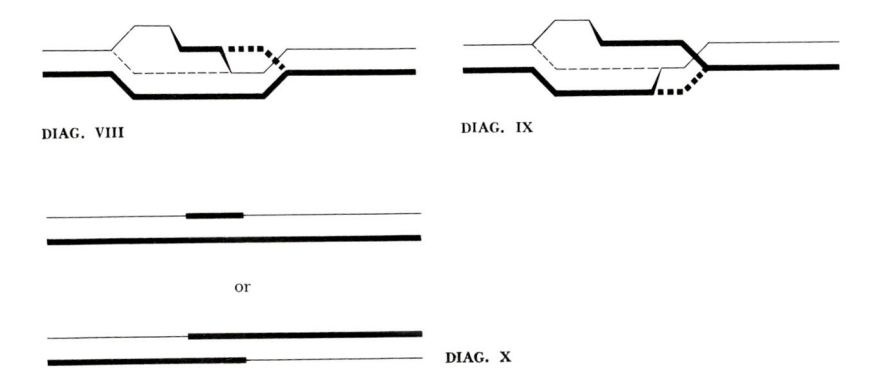

DIAG. VIII

DIAG. IX

or

DIAG. X

alright in our example above partly because I imposed a rule which I call "The Rule of Good Sense." It follows: At the occasion of the second exchange, the product of the nonreciprocal event which is saved is the one which permits the reconstruction of a continuous pair of chromatids. (I have no molecular mechanism in mind which could assure the operation of The Rule of Good Sense. A perennial failure to think of one would become an embarrassment to the model.) We can now examine the properties of the model by diagramming the consequences of a number of possible patterns of exchange. We shall relate the consequences of each pattern to the observed properties of recombination which are explained by them. In each case we shall examine only those exchange patterns which give rise to a + + spore pair.

DIGRESSIVE EXCHANGES IN SEX CIRCLES

When the second exchange involves the two arcs which were not involved in the first one, the left and right "wings" of the circles are exchanged. This exchange results in positive interference in each wing, so that we may neglect the occurrence of exchanges outside the circle. Thus, only three patterns are possible in the asci which produce a + + spore pair by virtue of digressive exchange. One exchange must, of course, fall between m_1 and m_2. The second exchange can be to the left of m_1, to the right of m_2, or between m_1 and m_2. The three patterns are shown in Diagrams XI to XIII. The relative frequencies

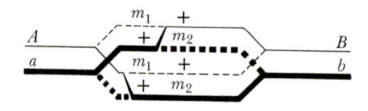

THE MECHANICS OF INHERITANCE

DIAG. XI

Resulting spore octet

$Am_1 + B \times 2$
$A + m_2b \times 2$
$a + + B \times 2$
$a + m_2b \times 2$

A 6:2 aberrant segregation at m_1 with recombination of outside markers.

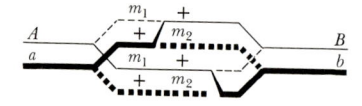

DIAG. XII

$Am_1 + B \times 2$
$Am_1 + b \times 2$
$a + + B \times 2$
$a + m_2b \times 2$

A 6:2 aberrant segregation at m_2 with recombination of outside markers.

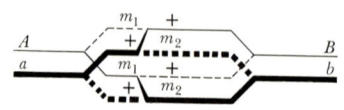

DIAG. XIII

$Am_1 + B \times 2$
$Am_1m_2b \times 2$
$a + + B \times 2$
$a + m_2b \times 2$

Reciprocal recombination of m_1m_2 with recombination of outside markers.

of XI and XII depend on the asymmetry of m_1 and m_2 in the circle. As drawn here, XII would be more frequent than XI. The frequency of XIII relative to XI plus XII depends on the distance between m_1 and m_2.

PROGRESSIVE EXCHANGES IN SEX CIRCLES

When the second exchange involves an arc which was involved in the first exchange, the left and right "wings" are not exchanged. No interference of crossovers in adjacent regions is set up, and we must consider cases involving additional exchanges outside of our circle (but presumably occurring in similar circles arising independently). The examples we give are illustrative, not exhaustive (Diagrams XIV and XV). Analogous patterns involving aberrant segregation of m_1 can be drawn. Again, the relative frequencies of aberrant segregation

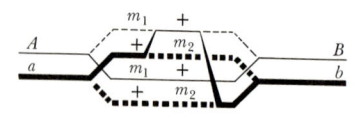

DIAG. XIV

Resulting spore octet

$Am_1 + B \times 2$
$Am_1 + B \times 2$
$a + + b \times 2$
$a + m_2b \times 2$

A 6:2 aberrant segregation of m_2 without recombination of outside markers.

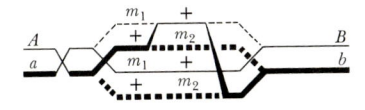

DIAG. XV

at the two loci depend on the asymmetry of m_1 and m_2 on the circle. The frequencies of the types involving three exchanges will depend on m_1–m_2 asymmetry as well. However, they will depend in addition upon the distances of A and B from the circle.

On page 173 we summarized the observations which our model was to explain. Make sure you understand how the model explains them! Other models to explain the observed correlations between aberrant segregation and crossing over have been devised. H. L. K. Whitehouse (1963, 1967) and R. Holliday (1964) indicate two of these. They rest in large part upon the idea of correction of base-pair mismatch in heteroduplex regions. As we described in Chap. 7, this idea is suggested by the behavior of enzymes which repair damaged duplex DNA. It is a good idea, and you might like to see just how each of those authors have developed it.

Summary

Aberrant segregation accompanies recombination of close markers in higher organisms. This is correlated with classical, reciprocal recombination of markers which bracket the aberrantly behaving ones. Complicated as it all seems, we must study it and think about it if we hope to understand crossing over.

References

Fogel, S., and D. D. Hurst, "Meiotic Gene Conversion in Yeast Tetrads and the Theory of Recombination," *Genetics, 57* (1967), 455–81. An outstanding example of the power of tetrad analysis.

Holliday, R., "A Mechanism for Gene Conversion in Fungi," *Genet. Research, 5* (1964), 282–304. A theory for crossing over and aberrant segregation based on the idea that heteroduplex heterozygosity stimulates the removal and resynthesis of stretches of DNA.

Meselson, M. S., "Reciprocal Recombination in Prophage Lambda," *J. Cellular Physiol., 70* (1967), Suppl. 1, 113–18.

Murray, N., "Polarized Intragenic Recombination in Chromosome Rearrangements of *Neurospora*," *Genetics, 58* (1968), 181–91. A good recent entrée to the literature of fixed pairing segments.

Whitehouse, H. L. K., "A Theory of Crossing-over by Means of Hybrid Deoxyribonucleic Acid." *Nature, 199* (1963), 1034–40. A model for aberrant segregation and crossing over.

Whitehouse, H. L. K., "Secondary Crossing-over." *Nature, 215* (1967), 1352–59. An attempt to bring the earlier model into line with recent data.

Problems

9.1. Let's pose some problems in terms of the cross $Am_1+B \times a+m_2b$. Call the intervals defined by the markers interval I, II, and III respectively, reading from left to right. The three intervals defined by the markers give the following frequencies of recombinants in two-factor crosses:

Cross	Frequency of recombinants
$Am_1 \times a+$	0.10
$m_1+ \times +m_2$	1.0×10^{-4}
$+B \times m_2b$	0.10

In the four-factor cross, the rare $++$ spores (5×10^{-5} of all the spores) were selected and scored for markers at loci A and B. The following numbers of each type were observed:

Type	Number
AB	320
ab	80
aB	510
Ab	90
Total	1,000

(a) Calculate the coefficient of coincidence for simultaneous recombination in regions I and II.
(b) Calculate the coefficient of coincidence for simultaneous recombination in regions II and III.
(c) In terms of a fixed pairing segment model, are m_1 and m_2 disposed toward the right or toward the left of the segment?
(d) In terms of our sex-circle model, which of the two loci (m_1 or m_2) shows the higher rate of aberrant segregation?
(e) In the sex-circle model, suppose that progressive and digressive exchanges between arcs were not equally probable. Which class of exchanges would have to be favored to account for an observed excess of parental types AB and ab over recombinant types aB and Ab among $++$ individuals?

9.2. Try to find a simple modification of the sex-circle model which will insure the operation of The Rule of Good Sense. (Drop me a line when you find one.)

9.3. In accounting for postmeiotic segregation, the sex-circle model assumed Watson–Crick duplication of heteroduplexes. How might our views of aberrant segregation change if we supposed that DNA duplicates in the manner proposed by Jehle?

Ten

The Code

In this volume we have dealt with the central questions of "transmission genetics"—how is the genic material duplicated, packaged, and partitioned? What is the structure of the genic material, and how is it mutated and recombined? By and large, we have avoided confrontation with questions most directly related to "physiological genetics," for example, the questions of how the genic material directs the metabolism and development of the creatures in which it resides. Fortunately other volumes in this series are devoted to that question, which is too big and has answers too rich to digest here. However, much of what we now know about the primary structure (the nucleotide sequence) of DNA has derived from studies on the mehcanisms by which the genic material directs protein synthesis. This chapter will describe the experiments which established the general rules relating nucleotide sequences of DNA to amino acid sequences of peptides synthesized under its direction (that is, the general nature of the genetic code). In addition it will describe, without presenting evidence, our present picture of the mechanism of protein biosynthesis. With this background, the experiments which cracked the code can be understood. Armed with the code, geneticists have been been able to look at problems of interest to "transmission genetics" not previously accessible.

The idea that things might be simple is a useful one.

It promotes simple hypotheses, which lead to simple, and consequently relatively decisive, experiments. The idea that there exists a simple relationship between the sequence of nucleotides in DNA and the sequence of amino acids in the proteins produced under its direction is a fine example. It provided the conceptual framework for the activities of geneticists and biochemists which not only resulted in code-cracking but also contributed centrally to our entire present-day understanding of the mechanism of protein synthesis.

The Mechanism of Peptide Synthesis—an Outline

Let us confine our attention to a stretch of DNA responsible for the amino acid sequence of a single polypeptide, that is, to a cistron. The first step in the synthesis of the peptide is the production of an RNA molecule the nucleotide sequence of which is usually a faithful reflection of the nucleotide sequence of the DNA. This RNA molecule, which is called a "messenger RNA" (mRNA) for reasons which may become apparent, is produced by an enzymatically catalyzed polymerization of nucleoside triphosphates. The *sequence* of the monomers is determined by base-pairing complementarity between *one* of the two DNA chains and the growing mRNA chain. The production of RNA on a DNA template may be usefully likened to the production of a new DNA chain on a complementary DNA template; you will recall from Chap. 2 that RNA is chemically similar to DNA, containing ribose instead of deoxyribose and uracil instead of thymine. The production of mRNA off a stretch of double-strand DNA may be pictured as in Fig. 10.1. The synthesis is stepwise, and the product mRNA contains information which is colinear with that of the DNA segment from which it was copied. "Transcription" is the word used to refer to the production of an RNA molecule on a DNA template. (The information in the DNA is "transcribed" into RNA.) The second step in polypeptide synthesis is the "translation" of the nucleotide sequence of mRNA into the amino acid sequence of the corresponding polypeptide.

Translation, like transcription, proceeds by the stepwise growth of a polymer, the polypeptide. Translation is catalyzed by small organelles (about the size of a virus) called ribosomes. A ribosome binds to an mRNA molecule, the mRNA is moved longitudinally across (through?) the ribosome presenting its information bit by bit, and a peptide is built bit by bit based on this information. The translation process exploits complementary base-pairing (as do DNA and RNA duplication and mRNA production). The individual amino acids are prepared for polymerization by becoming bonded to small RNA molecules. For each amino acid there exists one, or a few, species of such molecules; the ap-

The Code

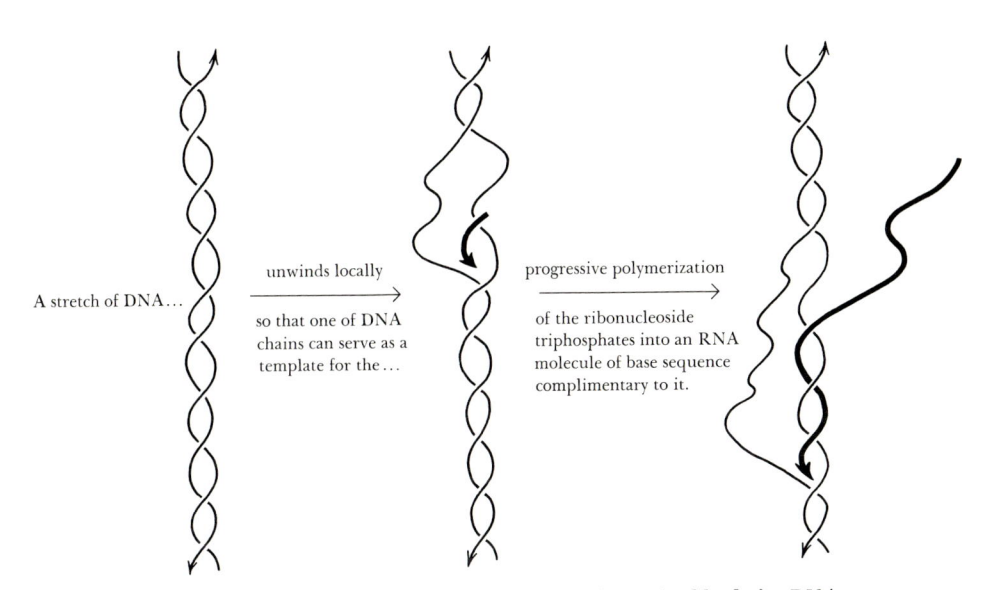

FIG. 10.1. *Scheme of RNA synthesis directed by a stretch of double-chain DNA. The arrowheads indicate chemical polarity of the polynucleotide chains. The head of the arrow is at the end of the chain bearing a 3' —OH group. Nucleotides are added sequentially to the growing end of the chain in an enzymatically catalyzed event similar to that (diagrammed in Fig. 3.5) occurring during nucleic acid duplication.*

propriate union is assured by a set of enzymes. For each amino acid the cell contains a species of enzyme which catalyzes the union of that amino acid with its corresponding species of small RNA molecule. The small RNA molecules are called sRNA (because they are relatively *soluble*) or tRNA (because they *transfer* amino acids to the ribosome-mRNA complex for polymerization). Each tRNA molecule (about 70 nucleotides long) contains a short stretch of nucleotides complementary to the sequence of nucleotides on the mRNA molecule which "stands for" the corresponding amino acid. The tRNA molecules pair by H-bonding with these sequences (codons) giving up their amino acids to the polypeptide under construction as they do so. The events involved in the addition of an amino acid to a growing polypeptide chain are diagrammed in Fig. 10.2.

The picture of polypeptide synthesis outlined above implies a set of formal relations between DNA, RNA, and polypeptides. The information content of DNA (its particular nucleotide sequence) is transcribed into RNA of corresponding sequence; this RNA is in turn translated into a sequence of amino acids. Let us look at the rules governing transcription and translation (in so far as we know them) and the experiments which established them.

184 THE MECHANICS OF INHERITANCE

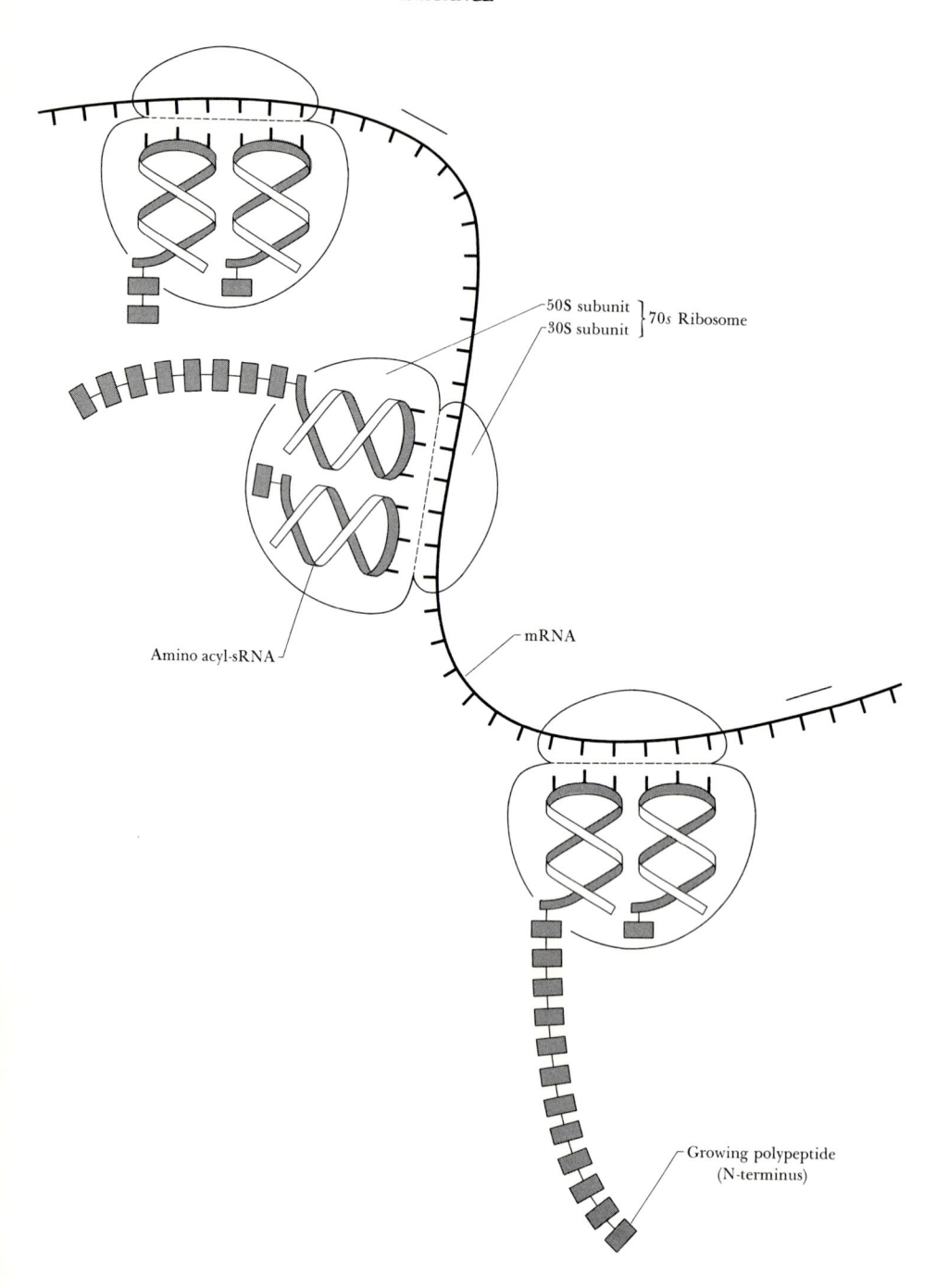

50S subunit ⎱ 70s Ribosome
30S subunit ⎰

Amino acyl-sRNA

mRNA

Growing polypeptide
(N-terminus)

FIG. 10.2. *Each ribosome has two sites which bind tRNA. One of the sites binds the tRNA molecule to which the growing peptide is attached; the other binds the tRNA molecule to which is attached the amino acid to be added. The proper selection of tRNA at the second site is assured by the base sequence of the bit of mRNA molecule (codon) which is presented at the second site. The growing peptide chain is then removed from the tRNA molecule to which it was attached and linked via peptide bond formation to the incoming amino acid. The tRNA molecule which lost its peptide chain is then displaced from its site by the molecule which now carries the chain as the messenger moves one codon's length across the ribosome. The vacated ribosome site now binds the next tRNA-amino acid molecule as dictated by the mRNA codon which has moved into that site. The chemistry of these steps in protein synthesis is presented in detail in Hartman and Suskind's* Gene Action, *a volume in this series. From I. H. Herskowitz,* Basic Principles of Molecular Genetics *(Boston: Little, Brown & Company, 1967).*

RULES FOR TRANSCRIPTION

(1) For any cistron, only one of the two complementary DNA chains is transcribed (see Fig. 10.1). As we set down further rules, you will be able to assess the confusion which would result were both chains of a given cistron transcribed. At this point we shall simply support the rule by examining evidence obtained from a creature particularly suited to the role. In Chap. 3, we met viruses in which the DNA of mature virus particle was single stranded. For some of these viruses (for example, the phage ϕx174) it has been shown that the infecting single-strand DNA directs the synthesis of a complementary DNA strand (becoming thereby double-strand) promptly after entry into the host bacterium. Soon thereafter, mRNA is synthesized which has nucleotide sequences complementary [1] to the newly produced DNA strand. None of the RNA produced is complementary to the infecting DNA strand.

(2) Transcription of a stretch of DNA begins at one end of a cistron. For many cistrons, transcription is terminated at the other end. For some cistrons, however, transcription proceeds past that end and continues until an mRNA molecule is made which is a transcript of several (adjacent) cistrons. Some of the evidence for such multicistronic

---

[1] Annealing of nucleic acids is used to determine the degree of complementarity. Heating disrupts the structure of double-chain DNA, permitting separation of its single chains. If such a heated suspension is rapidly cooled, more or less complementary regions between (or within) chains become "frozen" into imperfect duplex structures. If the suspension is slowly cooled (annealed) however, such metastable configurations do not get frozen in. Under the best conditions, the original duplex structure of heated DNA can be totally restored. The extent of complementarity between two preparations of nucleic acid can be determined by the extent to which molecules in the two preparations will anneal with each other.

messengers is ingeniously indirect; it is described in Hartman and Suskind's *Gene Action,* in this series. Our present purposes, however, are served by noting that the high molecular weight of some mRNA molecules indicates their multicistronic nature. The quantitative implications of this statement can be deduced after we have examined the rules governing translation.

(3) Messenger RNA chains "grow" by the successive addition of 5'-ribonucleotides at the 3'—OH group of the growing chain ("RNA is synthesized in the 5' to 3' direction"). Analysis of in vitro enzymatic synthesis of RNA on DNA templates provides the most direct evidence for this contention, as it did for the case of DNA synthesis on DNA templates or RNA on RNA templates as described in Chap. 3.

RULES FOR TRANSLATION

(1) Translation proceeds by the stepwise addition of amino acids to the growing polypeptide chain. A peptide chain has a free amino group at one end and a free carboxyl at the other (see Fig. 1.5). The stepwise synthesis proceeds from the amino to the carboxyl end of the chain. In the first experiment to establish the direction of protein synthesis, H. Dintzes fed radioactive amino acids to immature red blood cells whose protein-synthesizing machinery is devoted to hemoglobin synthesis. He examined completed protein molecules shortly after the radioactive amino acids had been added and found that the radioactivity incorporated into protein molecules was exclusively located close to the carboxyl-terminal ends. Samples examined at later times were labeled with radioactive amino acids in a more nearly uniform fashion.

(2) The mRNA is translated proceeding from the 5' to the 3' end of the RNA molecule.

After you learn the details of the code you may be able to think of an experiment which tests this contention.

(3) Translation begins at a certain position on the mRNA molecule, and successive sequences of three nucleotides each (triplets, codons) serve to direct the incorporation of amino acids. The several implications of this rule are best appreciated by examining in detail the experiments which secured it.

Frame-Shift Mutations and the General Nature of the Genetic Code

In the early 1960s, F. Crick and associates put the B cistron of Benzer's *rII* region to work in a series of experiments which established the grammatical rules relating the language of nucleic acid to that of protein. It was becoming apparent from studies on the induction of mutants (such as those described in Chap. 4) that most (about 80 percent) of the mutations occurring spontaneously in the *rII* region were not base-pair transitions. These mutants resembled those which arise during growth of T4 in the presence of acridine dyes. Acridine-induced mutants cannot be made to revert by base analogues or by nitrous acid, but they are spontaneously revertible, and they can be induced to revert by acridine dyes. (Analogue-induced mutations are not revertible by acridines, either.)

The most convenient measure of reversion for *rII* mutants is the restoration of the ability of the phage to make a plaque on *E. coli* strains harboring prophage λ. It is occasionally the case that a clone which meets this criterion of being revertant gives *r*-like rather than wild-type plaques when plated on *coli* strain B or B (λ). For some of the nontransition *rII* mutants such "pseudo-wilds" are the rule. When these pseudo-wilds are crossed to wild-type T4, two types of recombinant progeny arise at low (0.1 percent or less) frequency. One recombinant is the original *r* mutant, and the other is also an *r* mutant, which, like the first, is unable to grow on *E. coli* (λ). In addition, the second mutation is like the first in that it is not induced to revert by base-analogues or by nitrous acid, but only by acridines; it is an "acridine-type mutation." For a single acridine-type mutation in the *rII*B cistron, Crick *et al.* were able to isolate a number of different mutations which could restore its ability to grow on *coli* (λ); a number of acridine-type mutants were collected by repeating this procedure. The mutants were then combined (by genetic recombination) into many of the possible pairwise combinations. Some of the combinations were pseudo-wild and the others were fully *rII*, that is, unable to grow on *coli* (λ). Depending on the phenotype of the double mutant, mutants could be put into two classes, + and −, defined by the following rule: two mutants belonging to the same class never [2] offset each other's mutant effects; two mutants belonging to different classes may offset each other's effects. The guess as to the nature of the acridine-type mutations depended upon one other observation—these mutations were never found in cistrons whose protein products were essential in the life cycle of the phage. From this observation it could be concluded that the acridine-type mutations profoundly altered the nature of the protein product of the cistron in which they occurred. However, from the revertibility of these mutations it could be concluded that they were not large deletions or complex rearrangements of nucleotide sequences. The ability of the members of some pairs to compensate for each other's effects argues similarly, as does the small extent of these mutations as judged from fine-structure mapping experiments.

Crick offered the following explanation for the paradoxical acridine-type mutants: (1) mRNA is translated into protein one amino acid at a time, beginning at a start signal which sets the translation mechanism in the proper reading frame. In the normal course of events, successive

[2] Well, hardly ever.

188 THE MECHANICS OF INHERITANCE

codons (of uniform length) are properly viewed and translated by the precise shifting of the translation machinery exactly one codon's length along the messenger along with the addition of each amino acid to the growing polypeptide chain (see Fig. 10.2). (2) Acridine-type mutations are small deletions and/or additions of nucleotides. If the number of nucleotides deleted or added is not an integral multiple of the number of nucleotides in a codon, the result is a shift of the reading frame ("frame-shift" mutations) and consequently mistranslation of the portion of the subsequent message. Frame-shift mutations of opposite sign may compensate because they produce compensating shifts in the reading frame. (Compensation of the reading frame-shifts is a necessary, though insufficient, condition for mutual compensation of mutant effects of two frame-shift mutants. The additional requirement is that the message between the two mutant sites codes for a more or less satisfactory sequence of amino acids.) Recently, the amino acid sequences of several proteins produced by cistrons bearing compensating frame-shift mutations have been examined. The pseudo-wild proteins differed from the wild-type protein over a stretch of several amino acids, as predicted by Crick's hypothesis.

The amino acid sequence determinations also demonstrated that some frame-shift mutations are additions, some are deletions, and some involve the loss or addition of more than one nucleotide. Before presenting the evidence, I would like you to convince yourself that mutations involving the addition of two nucleotides are in the same class (that is, affect the reading frame in the same way) as deletions of one nucleotide. Similarly, of course, deletions of two nucleotides are in the same class as additions of one nucleotide. The evidence is the demonstration of pseudo-wild proteins which are one amino acid shorter than the true wild-type protein and of other pseudo-wild proteins which are the same length as the wild-type protein.

Without waiting for confirmation of their hypothesis via the examination of pseudo-wild proteins, Crick et al. parlayed their notion into a determination of the number of nucleotides in a codon. They reasoned that n nearby frame-shift mutations of the *same* sign could result in a pseudo-wild type if n is the number of nucleotides in a codon. Make sure you see why! You will then understand the extent to which Crick's demonstration that the combination of three frame-shift mutations of the same sign can produce pseudo-wilds argues for a triplet code (three nucleotides per codon). The case was strengthened by noting that six (but not four or five) mutations of the same sign can also combine to give pseudo-wilds.

Cracking the Code

Proteins can be synthesized in vitro by a mixture of ribosomes, tRNA with amino acids attached (or tRNA plus amino acids plus the enzymes and ATP needed to make the attachment), mRNA, and a few other compounds whose roles are not clear. In the early 1960s, M. Nirenberg added polyuridylic acid (synthetic RNA whose only base was uracil) in place of mRNA in such a mixture. The polyuridylic acid directed the synthesis of a polypeptide containing only phenylalanine. The prompt and correct conclusion was that the codon for phenylalanine is a triplet of uracils (UUU). The way was paved for an assault on the code which led to its full solution by 1966. The resulting dictionary is given in Table 10.1.

TABLE 10.1. *mRNA Codons for Amino Acids*

Second base

	U	C	A	G	
U	UUU, UUC } phe UUA, UUG } leu	UCU, UCC, UCA, UCG } ser	UAU, UAC } tyr UAA non UAG non	UGU, UGC } cys UGA non UGG try	U C A G
C	CUU, CUC, CUA, CUG } leu	CCU, CCC, CCA, CCG } pro	CAU, CAC } his CAA, CAG } gln	CGU, CGC, CGA, CGG } arg	U C A G
A	AUU, AUC, AUA } ileu AUG met	ACU, ACC, ACA, ACG } thr	AAU, AAC } asn AAA, AAG } lys	AGU, AGC } ser AGA, AGG } arg	U C A G
G	GUU, GUC, GUA, GUG } val	GCU, GCC, GCA, GCG } ala	GAU, GAC } asp GAA, GAG } glu	GGU, GGC, GGA, GGG } gly	U C A G

First base (left); *Third base* (right)

U, C, A, and G stand for the four bases in RNA. For each triplet the left end is the 5' end, that is, the triplets are "read" from left to right. The abbreviations for the amino acids are given in Fig. 1.4b; *non* is short for "nonsense." The three nonsense triplets code for no amino acid (in most creatures). Their presence in a message leads to termination of peptide chain growth at that point. Perhaps one or another of them, or some combination of them, is part or all of the normal signal to terminate peptide chain synthesis at the end of a cistron.

Knowledge of the code provided a new avenue to the study of mutation. The direct determination of nucleotide sequences of DNA is not yet possible (although one is encouraged by the recent determination of the nucleotide sequences of several types of tRNA). However, amino acid sequences of proteins can be determined, though it takes a lot of work for each. Knowledge of an amino acid sequence provides partial knowledge of the nucleotide sequence which generated it. Thus, comparisons of the amino acid sequences of mutant and wild-type proteins can sometimes reveal the exact nature of a mutational event. For instance, if a mutant protein contains *phe* at the position normally occupied by *ser,* the mutational event must have been a transition of a CG base pair to a TA pair (at the second position of the codon). Evidence for transversions has been provided by such analyses; amino acid differences between mutant and wild-type proteins which could only have arisen by transversion have been described. In fact, some of the "mutator" strains described in Chap. 4 produce primarily transversion mutations.

Knowledge of the code has already provided clues to the mechanism of frame-shift mutation, too. These mutations occur at the highest rates in regions which are repetitious with respect to base sequence. In the case of frame-shift mutations which add nucleotide(s), the added nucleotide(s) tend to be the repetitive ones. These observations by Streisinger led him to the following theory of frame-shift mutagenesis: in double-strand DNA, regions near a chain end are free to melt and refreeze in new, metastable configurations. Thus, the intermediates in recombinant formation in phages (see Chap. 8), which look like this,

can melt and refreeze like this,

$$\begin{array}{c}(T)\\ T\;T\;T\;C\;G\;T\;A\;G\;C\\ A\;A\;A\;A\;G\;C\;A\;T\;C\;G\end{array}$$

so that sealing of the gaps leads to,

$$\begin{array}{c}(T)\\ T\;T\;T\;T\;C\;G\;T\;A\;G\;C\\ A\;A\;A\;A\;G\;C\;A\;T\;C\;G\end{array}$$

This heteroduplex is frame-shift mutant on the upper chain. Upon its duplication, pure mutant molecules (and wild-type molecules) will arise. Now see whether you can play the same game of slipped pairing near chain ends to produce frame-shift deletions. If you succeed, you will understand why addition mutants like the one diagrammed on the preceding page have especially high reversion rates (and why these revertants are usually true wild-type). While you are thinking, propose a mechanism whereby acridines, which are known to interact with DNA such as to increase the stability of the duplex, promote the production of frame-shift mutations.

Streisinger's theory provides a good (that is, testable) hypothesis for the variation from one position to another on the chromosomes in the rate of occurrence of frame-shift mutations. The transitions and transversions, too, vary in rate from one site to another according to the neighboring sequences of bases. Thanks to our knowledge of the code plus our ability to determine amino acid sequences, we should soon know which nucleotide sequences promote base-substitution mutagenesis. The fundamental problems of genetics which have become accessible via the code are being daily identified—and solved. You will have to hurry if you want to help!

References

The Genetic Code (Cold Spring Harbor Symp. Quant. Biol., 31), Cold Spring Harbor Laboratory of Quantitative Biology, Cold Spring Harbor, New York (1966). Judicious skimming will introduce you to the ideas, the personalities, and the original literature which were involved in the cracking of the code.

Problems

10.1. On Mars, life is based on protein and nucleic acid, as is our own. However, the proteins are made of 312 different amino acids instead of 20, and there are 6 kinds of nucleotides instead of 4. Otherwise, things are pretty much the same. This information permits you to set a lower limit to the number of nucleotides in a Martian codon. At least how long must a codon be up there?

10.2. From the codons for which amino acids can one derive codons for lysine by single base-pair transitions? By single base-pair transversions?

10.3. A creature whose DNA was double-stranded produced an mRNA molecule whose nucleotide sequence begins as shown:

Nucleotide: A U G A C A U A U C A U A G G C C C U U U G G G ...
Position
No.:   1 2 3 4 5 6 7 8 9 1 1 1 1 1 1 1 1 1 1 2 2 2 2 2
                               0 1 2 3 4 5 6 7 8 9 0 1 2 3 4

(a) Write the amino acid sequence for the NH_2-terminal end of the polypeptide chain whose synthesis is directed by the messenger above.
(b) Write the amino acid sequence for the mutant which has been deleted at position No. 6.
(c) Write the amino acid sequence for the double mutant which has been deleted at positions 8 and 9 and position 22.
(d) The double mutant in (c) was grown on a host in which the nonsense triplet UAG is translated as serine (*ser*). Write the amino acid sequence of the polypeptide produced.

192 THE MECHANICS OF INHERITANCE

(e) The double mutant in (c) was treated with nitrous acid. Two kinds of mutants (pseudo-wilds) were induced in which polypeptide chains of normal length were produced. Write their amino acid sequences.

10.4. The sequence of amino acids in a particular region of an enzyme made by an Earth creature is ... *arg-leu-val-thr-gly-ser-try-leu-tyr*.... A mutant was isolated which made no active enzyme. Some pseudo-wild revertants of that strain were isolated which made partially active enzyme. The enzyme of one of these had the following amino acid sequence: ... *arg-leu-ser-gln-val-his-leu-tyr*.... On the basis of this information, specify as many as possible of the nucleotides in the mRNA specifying the wild-type enzyme. (You should be able to specify 18 nucleotides completely and 6 of the remaining 9 partially.)

10.5. The comparison of physical distances with map distances is facilitated by our knowledge of the code; the code permits us to "translate" numbers of amino acids in a protein into numbers of nucleotides on the nucleic acid.

(a) Consider two mutant forms of a particular enzyme in *Neurospora*. The two proteins are known to differ only at two positions in their polypeptide chains. The two positions are separated by 40 other amino acids. Crosses of the two mutants regularly give 1.0×10^{-6} recombinants. Approximately how many nucleotides in *Neurospora* separate a pair of markers which gives 10^{-5} recombinants?

(b) Consider two other mutant forms of the enzyme. Each has the normal (wild-type) number of amino acids, but they differ from each other and from the wild-type enzyme at position number 68 in the polypeptide chain. The wild-type protein has argininine at position 68 whereas the mutants at Nos. 1 and 2 have glutamine and serine respectively. When mutants 1 and 2 were crossed, six kinds of offspring differing in the amino acid at position 68 were observed. Two of these were parental types in which glutamine and serine were present at position 68. Three novel types—having lysine, asparagine, and histidine, respectively—made up the remainder. The types and the frequencies at which they arose in the cross are tabulated here.

	Amino acid at position 68	Frequency among offspring of the cross
Parental types	gln	0.50
	ser	0.50
	arg	4.0×10^{-7}
	his	2.0×10^{-7}
	asn	1.0×10^{-11}
	lys	2.0×10^{-7}

First, deduce the nucleotide sequences at codon number 68 for each of the six types. (1) What is the frequency of recombinants for crosses in which the markers are at adjacent positions in the DNA? (2) What is the coefficient of coincidence for recombinations at adjacent inter-nucleotide links?

Eleven

The Genetic Analysis of Diploids

This last chapter will serve several purposes. It will provide a partial review of ideas introduced in previous chapters. It will provide a bridge to Hartman and Suskind's *Gene Action* as well as to other volumes in this series. It will provide a point of departure for students who wish to read in the early literature of genetics.

Genotype and Phenotype

The genetic markers carried by an individual constitute its genotype. In the examples used to this point, the genotype of each individual was generally deducible from a glance at the appearance (the phenotype) of the individual. In this chapter we shall deal frequently with exceptions to this situation.

Phenotypic Lag and Phenotypic Mixing

The distinction between genotype and phenotype is conveniently emphasized by one phenomenon occurring in bacteria and another occurring among phages.

When a mutation occurs in a multiplying bacterial population, several generations may pass before individuals of mutant phenotype appear. Two sources of this phenotypic lag can be easily identified for the case of a mutation from prototrophy to auxotrophy. (1) A bacterial cell that mutates to the inability to produce a certain enzyme may give rise for several generations to progeny cells that contain enzyme molecules partitioned out to them at the time of cell division. Such cells will manifest nutritional properties like those of their prototrophic ancestor, even though essentially all of their descendants are destined to be auxotrophic. These cells are genotypically auxotrophic but have a prototrophic phenotype. (2) Under some conditions of growth, many strains of bacteria are multinucleate; each cell possesses more than one set of genic material (see Fig. 5.6). Clearly, the phenotypic manifestation of a mutation to auxotrophy in one of these chromosomes must await the partitioning of that chromosome into a cell that contains no nonmutant chromosomes. Please note that our analysis of mutation in bacterial populations in Chap. 1 ignored phenotypic lag.

Mutants of phages are known that produce altered structural components of their protein coats. When mutant and wild-type phages are grown in the same bacterial cell, the emerging particles possess protein coats the components of which are drawn without regard to the genotype of the particle from a pool of mutant and wild-type protein molecules. Particles with such mixed coats may manifest some properties characteristic of one or the other (or both) phage strains while other properties may be intermediate between those of the two strains. Phenotypic mixing has been detected for stability at high temperatures, serological properties, and for the ability of particles to attach to different strains of bacteria, as well as for other properties that depend on the protein moiety of mature phage particles.

Dominance and Recessiveness

The phenotype of a phage-infected cell is a reflection of the genotype of the particle that initiates the infection. For instance, *rII* mutants of T4 fail to complete an infectious cycle when they infect λ-carrying strains of *E. coli*. Wild-type (*rII+*) phages, on the other hand, can develop normally in such *coli* strains. What is the phenotype of a cell infected with one particle of each genotype? Such infected cells have a wild-type phenotype; they produce a full crop of phage particles (half of which have the *rII* genotype and half of which are *rII+*). Under these conditions the presence of the *rII+* determines the phenotype; the *rII+* marker is said to be dominant relative to its recessive *rII* allele. The dominance of *rII+* is not absolute. If a single *rII+* particle infects *coli* (λ) along with *several rII* particles, the infected cell often *fails* to produce phages.

Nature and Nurture

The phenotype of an individual is a function not only of its genotype but also of the environment in which the creature develops. Phages of the *rII* genotype have a *lethal* phenotype when developing in *coli* (λ), but they develop perfectly well in *coli* cells that do not harbor λ. The lethal phenotype of *rII* mutants in *coli* (λ) is itself subject to further environmental modification. If the infection proceeds in a medium of low sodium ion concentration the *rII* phages develop normally.

Complementation

In Chap. 7, we discussed complementation between different *rII* mutations in T4. By definition, those *rII* mutations which were in separate cistrons could complement each other while those in the same cistron could not. (Interesting violations of both of these aspects of the definition of a cistron are described in Hartman and Suskind's *Gene Action*.) The functions of the two *rII* cistrons are clearly closely related. A more general illustration of complementation would be one in which the complementing mutants are not tightly linked and are not concerned with obviously related functions.

Mutants of T4 are known that can complete every step in their life cycle except for the "last" one. These mutants (*e* mutants) fail to produce a sufficiently active form of the enzyme (lysozyme) that digests the wall of the host cell permitting the escape of the particles. (Such mutants can be propagated by the geneticist who adds lysozyme to the growth medium.) Cells jointly infected by *e* and *e*+ particles lyse normally as a result of the action of the wild-type enzyme produced by the chromosome of *e*+ *genotype*. (With respect to the phenotype "cellular lysis," *e*+ is dominant over *e*. Partial dominance may be apparent at the molecular level; the infected cell may contain both wild-type lysozyme and the defective lysozyme characteristically produced in cells infected by the *e* mutant only.) *Escherichia coli* (λ) infected by *either rIIe*+ or *rII*+*e* particles fail to release phages. Cells infected simultaneously by particles of these two genotypes, however, produce and release a full crop of phages; the two genotypes complement each other.

Heterozygosis in Bacteria

In Chap. 8, we described circumstances in which large parts of a *coli* chromosome transferred during conjugation can coexist in the recipient cell with the original chromosome. The phenotype of such partial diploids illustrates in the expected fashion the phenomena of dominance, recessiveness, and complementation.

These partial diploids are heterozygous for those loci at which the donor and recipient cells carried different alleles. They are quite comparable to the diploid cells (heterozygotes) that arise by union of genetically different gametes in higher organisms.

Diploidy in Higher Organisms

In essentially all of the higher organisms, the formation of completely diploid cells is a regular feature of the life cycle. In some of these (for example, *Chlamydomonas*), the diplophase is transient and unexpressive; its phenotype is to an appreciable extent independent of its *particular* genotype. For the majority of the higher organisms, however, the diplophase is more persistent. Among the more familiar creatures (cats, firs, *Drosophila*) the diplophase is the conspicuous stage in the life cycle.[1] In such organisms, however, the genotypes of diploid individuals and their haploid meiotic products can generally be deduced from hybridization experiments. The rules for the conduct of good hybridization experiments were set by G. Mendel in the 1860s.

(1) Hereditary characteristics that can be scored with minimal ambiguity should be selected for study.

(2) For each locus of potential interest in the strains to be analyzed, the likelihood that the same marker is carried on each of the homologous Chromosomes should be maximized. This homozygosity can be made likely by employing individuals descended from a large number of generations of matings between phenotypically similar siblings (that is, by inbreeding). The effect of inbreeding on homozygosity is described in detail in Brewbaker's *Agricultural Genetics* in this series.

(3) The offspring of matings between individuals of two such inbred strains that differ from each other with respect to a few phenotypic characteristics should be selected for examination.

(4) Sibling matings with the hybrids from step three should be performed. Their offspring should be examined for the frequencies of occurrence of each of the possible phenotypes.

(5) It is often useful to mate the hybrids with each of the original inbred strains and examine the offspring of these backcrosses for the frequencies of occurrence of each of the possible phenotypes.

From just such experiments Mendel and the generation of geneticists

---

[1] In some creatures (yeast, for instance) both the diplophase and the haplophase can be propogated as single cells. The transition to diplophase can result from cellular fusion followed by nuclear fussion. Haploidization results from meiosis. This is an experimentally convenient situation, and the student ought to have no trouble reconstructing its genetic analysis.

that rediscovered his methodology in 1900 were able to reveal the phenomena of the union of gametes at random with respect to genotype, segregation, dominance and recessiveness, complementation, recombination (linked and unlinked), and interference (positive and negative), and to correlate these phenomena with the behavior of Chromosomes. The heroic nature of their triumph is apparent when we realize that the outcome of each one of their experiments was usually determined by the interplay of most of these phenomena, as well as some others. If I have been successful, my readers should find such analysis easy; I doubt that I've been so completely "successful" as to take the fun out of it. Try the problems, please, and we'll see.

Reference

Mendel, Gregor, "Experiments in Plant Hybridization," 1865. Reprinted in *Classic Papers in Genetics*, J. A. Peters, ed. (Englewood Cliffs, N.J.: Prentice-Hall, Inc., 1959), pp. 1-20.

Problems

11.1. Consider an organism which has an alteration of haploid and diploid generations in which the diplophase arises by nuclear fusion between two haploid cells (gametes), and the haploid cells are produced by meiosis. (Such is the case for cats, firs, and *Drosophila*.) Let's suppose that three genetic loci (1, 2, and 3) are known for this creature. Locus 1 can be occupied by marker *A* or its allele *a*; locus 2 can be occupied by *B* or its allele *b*; locus 3 can be occupied by *D* or *d*. Assume that the markers denoted by capital letters are dominant over their alleles, and that the three loci influence phenotype independently of each other. Now consider some diploid individuals all of which were formed by the union of one gamete of type *ABD* and the other of type *abd*. (The offspring of matings between two purebreeding strains are called collectively the F_1.)

(*a*) Considering just locus 1, what fraction of the haploid cells produced by meiosis in F_1 individuals will be of genotype *A*?

(*b*) Considering just locus 1, what genotypes of diploids can arise by random union of gametes produced by F_1 individuals? In what frequencies do the various genotypes arise?

(*c*) What fraction of the diploids produced in Problem 11.1*b* have the phenotype *a*?

(*d*) Considering just loci 1 and 2, what genotypes of gametes can be produced by F_1 individuals? In what frequency do the types occur if loci 1 and 2 are unlinked? If loci 1 and 2 recombine 20 percent of the time?

(*e*) Considering just loci 1 and 2, four possible phenotypes of diploids can arise by random union of gametes produced by F_1 individuals (call such offspring the F_2). We designate these as *AB*, *Ab*,

aB, and ab depending on whether they have the phenotype characteristic of the dominant or of the recessive allele at the respective loci. In what frequency do the various phenotypes arise in the F_2 if loci 1 and 2 are unlinked? If loci 1 and 2 recombine 20 percent of the time?

(*f*) Suppose that loci 1 and 2 recombine 20 percent of the time, that loci 2 and 3 recombine 10 percent of the time, and that loci 1 and 3 recombine more than 20 percent of the time. For the gametes produced by the F_1 write the genotypes that are recombinant simultaneously for each of the two pairs of neighboring loci. What are the frequencies of each of these double recombinants if the coefficient of coincidence (*S*) for the two adjacent regions of the map is 0.6? If *S* is 0?

11.2. Consider an agriculturally important plant heterozygous at four loci. It is known that three of the loci are on one Chromosome and that the other one is on a different Chromosome. The loci carry allelic pairs of markers determining height (short or tall) at locus 1, leaf color (red or green) at locus 2, leaf texture (rough or smooth) at locus 3, leaf width (narrow or wide) at locus 4. Suppose the heterozygous individuals arose from a cross of a pure-breeding (homozygous) strain that was tall, red, rough, and wide with another pure strain that was short, green, smooth, and narrow. In a test cross (a cross of an F_1 to a strain carrying only recessive alleles at all pertinent loci), the following progeny were produced in the indicated (invented) numbers:

Phenotype				Numbers of individuals
tall	red	rough	wide	128
short	green	smooth	narrow	134
tall	green	rough	wide	95
short	red	smooth	narrow	93
tall	red	smooth	wide	129
short	green	rough	narrow	133
tall	red	smooth	narrow	21
short	green	rough	wide	17
tall	green	smooth	wide	92
short	red	rough	narrow	96
tall	green	smooth	narrow	5
short	red	rough	wide	7
tall	red	rough	narrow	18
short	green	smooth	wide	20
tall	green	rough	narrow	6
short	red	smooth	wide	6
			Total	1,000

(*a*) Which locus is on a separate Chromosome?
(*b*) What is the order of the loci on a common Chromosome?
(*c*) What is the frequency of recombination for each of the adjacent loci in Problem 11.2*b*?

(d) What is the coefficient of coincidence for the neighboring regions of the map?

11.3. Suppose an inbred deep-purple pigeon-toed individual was mated with a white normal-toed one. All of their many offspring were violet-colored and normal-toed. When these offspring (the F_1) were mated to individuals of the inbred, deep-purple, pigeon-toed strain, offspring were produced in the following frequencies:

Phenotype	Frequency
deep-purple, pigeon-toed	0.40
violet-colored, normal-toed	0.40
violet-colored, pigeon-toed	0.10
deep-purple, normal-toed	0.10

It may help you to know that when the F_1 were mated to inbred individuals whose phenotype was white, pigeon-toed, they produced offspring with the frequencies shown below:

Phenotype	Frequency
violet-colored, pigeon-toed	0.40
white, normal-toed	0.40
white, pigeon-toed	0.10
violet-colored, normal-toed	0.10

(a) What is the recombination frequency for the two loci involved in this problem?
(b) When F_1 individuals are mated to each other, their offspring are called the F_2. What frequency of violet, normal-toed individuals would you expect to find among the F_2 of this problem? What frequency of white, pigeon-toed individuals?

11.4. Suppose a highly inbred tall, white, long-haired male individual was mated with an inbred short, red, hairless female. All of their offspring were tall, red, and long-haired. Siblings mated to each other produced the following kinds of offspring in the indicated proportions:

Phenotype	Frequency
tall, red, long-haired	0.5056
tall, white, long-haired	0.2394
short, red, long-haired	0.0046
tall, red, hairless	0.0042
short, white, long-haired	0.0002
short, red, hairless	0.2354
tall, white, hairless	0.0006
short, white, hairless	0.0096

(a) Calculate the frequency of recombination for each of the two map intervals involved.
(b) Calculate the coefficient of coincidence for the two map intervals.
(c) How might you redesign the experiment so as to extract the same information more easily?

11.5. Perhaps in some plant (such as corn) there exists an enzymatically catalyzed sequence of chemical reactions leading to the formation of a red substance Z. The precursors of Z are X and Y, both of which are colorless. The conversion of X to Y is accomplished by enzyme E_1, the conversion of Y to Z by enzyme E_2. Two different pure-breeding strains are known which fail to produce red pigment. The F_1 resulting from matings between individuals from the two different strains all produce red pigment. In the F_2 the ratio of red to nonred individuals was observed to be 0.5625 : 0.4375. What is the frequency of recombination between the two loci influencing the production of Z?

Appendix

The Poisson Distribution

Consider a sequence of N trials, the outcome of each of which is independent of the outcome of previous trials. Let the probability of success of any one trial be p and that of failure be $1 - p = q$. The probability P_n of exactly n successes in N trials is given by the $(n + 1)$ term of the expansion of the binomial $(p + q)^N$:

$$P_n = \frac{N!}{(N - n)!\, n!}\, p^n q^{N-n}$$

Many readers will know this already; others will be content to accept it. For those who fall into neither category, I offer a derivation of sorts.

Consider a bucket containing a vast number of marbles. A fraction p of the marbles are red, and the remaining fraction $q = 1 - p$ are green. Beside the bucket you have an array of N teacups numbered $1 - N$. Stir the contents of the bucket thoroughly, then close your eyes and grab a marble. Transfer the marble to one of the teacups. Now pick a second marble from the bucket and transfer that to a different one of the cups. Repeat the procedure until each of the teacups has a marble. *Our job is to calculate the probability that exactly n of the N teacups contain red marbles.*

The probability that any *one* cup you care to mention (without peeking) contains a red marble is p. The probability that any array of n cups you chose to designate contains only red marbles is p^n, since the marbles were selected blindly from a vast, well-mixed population. The chance that the *rest* of the cups (those you didn't designate) contain only green marbles is, by the same argument, q^{N-n}. Thus, the probability that any blindly designated array of n teacups contains only red marbles while the remainder contain only green is $p^n q^{N-n}$. If we can now figure out how many different arrays of n teacups can be designated, we've got our answer. That answer will be

$$P_n = p^n q^{N-n} \times \text{the number of different arrays of } n \text{ teacups that can be selected from an array of } N \text{ cups}$$

So, pick a cup! You have N choices available, and thus any one of N cups can be your choice. Whatever the outcome of that choice, there are $N - 1$ cups remaining, and any one of $N - 1$ cups can be your second choice. Therefore, the number of different sequences of n cups that can be drawn is

$$N(N-1)(N-2)\ldots[N-(n-1)].$$

Many of these sequences, however, may constitute identical arrays of n cups. The sequences shown below, for example, represent identical arrays of designated cups.

First choice	Second choice	Third choice
7	3	1
1	7	3
1	3	7
7	1	3
3	7	1
3	1	7

The number of different arrays of n cups that can be designated is given by

$$\frac{\text{Number of different sequences in which } n \text{ cups can be drawn from } N \text{ cups}}{\text{Number of different sequences in which } n \text{ cups can be arranged}}$$

In the example above, the first choice can be any one of three cups. Once the first choice is made, the second choice can be either of two. When that is fixed, the last choice is fixed, so that the number of different sequences is given by

$$3 \times 2 \times 1 = 3!$$

In general, for n cups, the number of different sequences is $n!$. The desired number of different arrays, then, is given by

$$\frac{N(N-1)(N-2)\ldots[N-(n-1)]}{n!}$$

You should have no trouble showing that this is equivalent to

$$\frac{N!}{(N-n)!\, n!}$$

Q.E.D.

The binomial distribution is the precise solution to our marble-pulling problem. You should use it when you can. However, there are numerous occasions in which an algebraic approximation to the binomial distribution is more useful. The approximation which we shall find especially useful is that of Poisson. Poisson's approximation to the binomial applies whenever N is very large but p is so small that the product Np is small compared to N. Under these conditions

$$P_n = \frac{(Np)^n e^{-Np}}{n!}$$

where e, the base of the natural logarithm, equals 2.718....

The mean of the Poisson Distribution (as well as of the binomial) is Np, so that we may write

$$P_n = \frac{x^n e^{-x}}{n!}$$

where x is the average number of successes in N trials.

In the problems which follow (as in all the problems in this volume), don't substitute numbers into *any* formula until you're sure that all the conditions for applying that formula are met.

References

Feller, William, *An Introduction to Probability Theory and Its Application*, Vol. I. New York: John Wiley and Sons, 1957. "A [beautiful] treatment of probability theory developed in terms of mathematical concepts." Quote from the dust jacket of the second edition with parenthetical comment by F. W. S.

Hodgman, Charles D., Mathematical tables from *Handbook of Chemistry and Physics*. Cleveland: Chemical Rubber Publishing Co. The mathematical tables of the *Handbook* will fit in your pocket. The sections on exponential functions and factorials are useful in problems involving the Poisson distribution, radioactive decay processes, mapping functions, and so on.

Problems

A.1. (a) Taking the value of e as 2.72, calculate the following to within an accuracy of a few percent: e^{-1}; e^{-2}; e^{-3}; e^{-4}; e^{-5}.
(b) In Fig. A.1, plot your values and draw a smooth curve through them.

(c) In Fig. A.2, plot your values again and draw a straight line through them.

(d) From the plot in Fig. A.2, determine the values of $e^{-1.5}$; $e^{-3.0}$; $e^{-4.3}$; $e^{-0.3}$; $e^{-0.1}$.

A.2. Consider a bucket containing 10 marbles, 7 of which are red, 3 of which are blue. What is the probability of each of the following samples?

(a) In a sample of size 2, both red.
(b) In a sample of size 2, 1 red and 1 blue.
(c) In a sample of size 5, 3 red and 2 blue.
(d) In a sample of size 5, 4 red and 1 blue.
(e) In a sample of size 9, 7 red and 2 blue.

A.3. (a) Consider a bucket containing an immense number of marbles, 70 percent of which are red, 30 percent of which are blue. What is the probability of each of the following samples? (1) In a sample of size 2, both red. (2) In a sample of size 5, 3 red and 2 blue. (3) In a sample of size N, n red and $N - n$ blue.

(b) Consider a bucket containing an immense number of marbles of which fractions 0.9999 are red and 0.0001 are blue, respectively. What is the probability of each of the following samples? (1) In a sample of size 1,000, the chance of 3 blues. (2) In a sample of size 10^5, the chance of 6 blues. (3) In a sample of size 10^4, the chance of no blues.

(c) Consider a bucket containing 2×10^{12} red marbles, 10^{12} blue marbles, and 10^{12} green marbles. Calculate the following probabilities: (1) In a sample of size 1 drawn randomly from the bucket, the probability of a green marble; of a red marble. (2) In a sample of size 2, the probability of a blue and a green; of a red and a blue. (3) In a sample of size 5, the probability of 3 reds. (4) In a sample of size 6, the probability of 2 greens.

(d) Suppose 10^4 purple marbles are tossed into the bucket in Problem A.3c. Calculate the following probabilities in a sample of size 4×10^8. (1) The probability of no purple marbles. (2) The probability of 3 purple marbles. (3) The probability of 2 or more purple marbles.

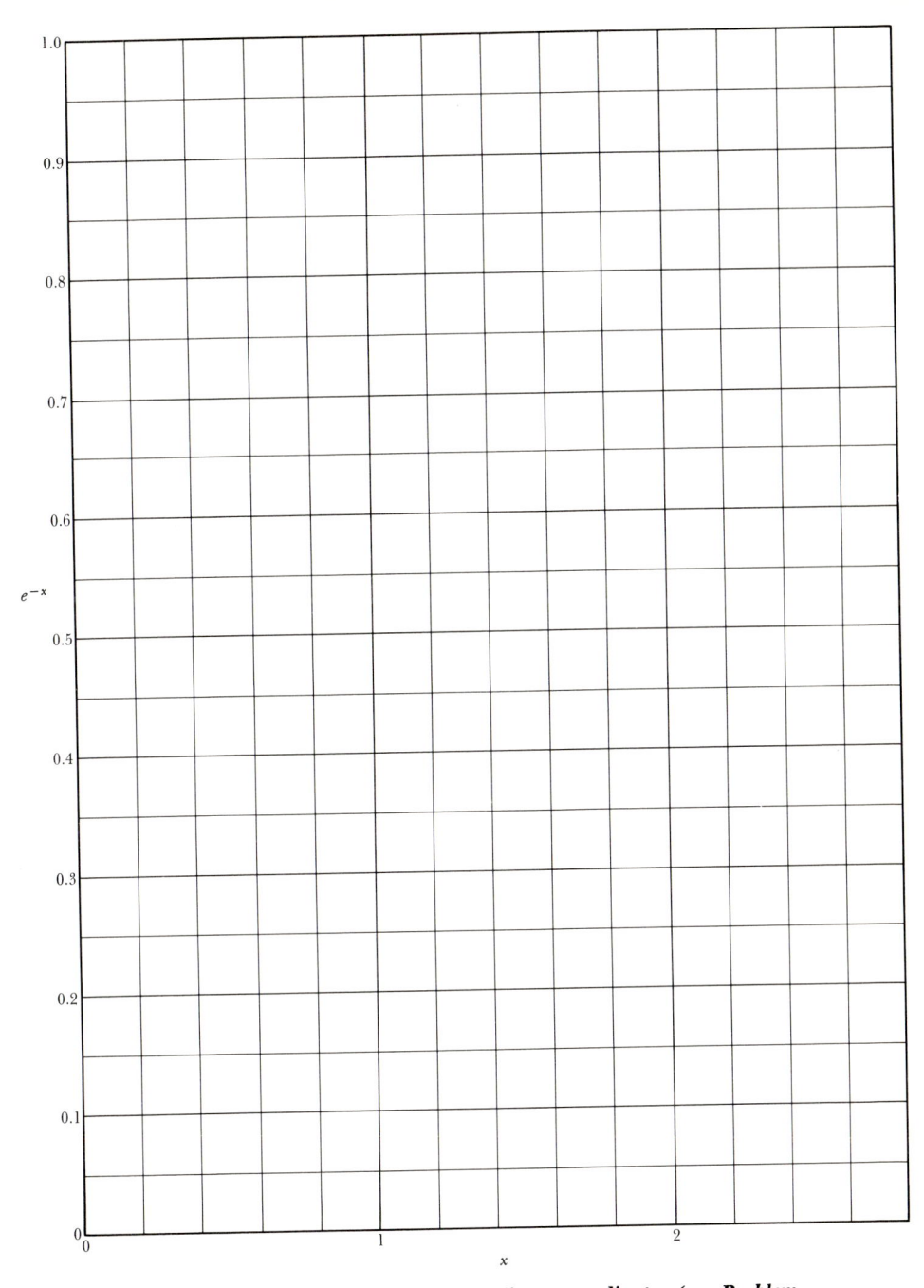

FIG. A.1. *A do-it-yourself plot of* e^{-x} *versus* x *on linear coordinates (see Problem A.1b).*

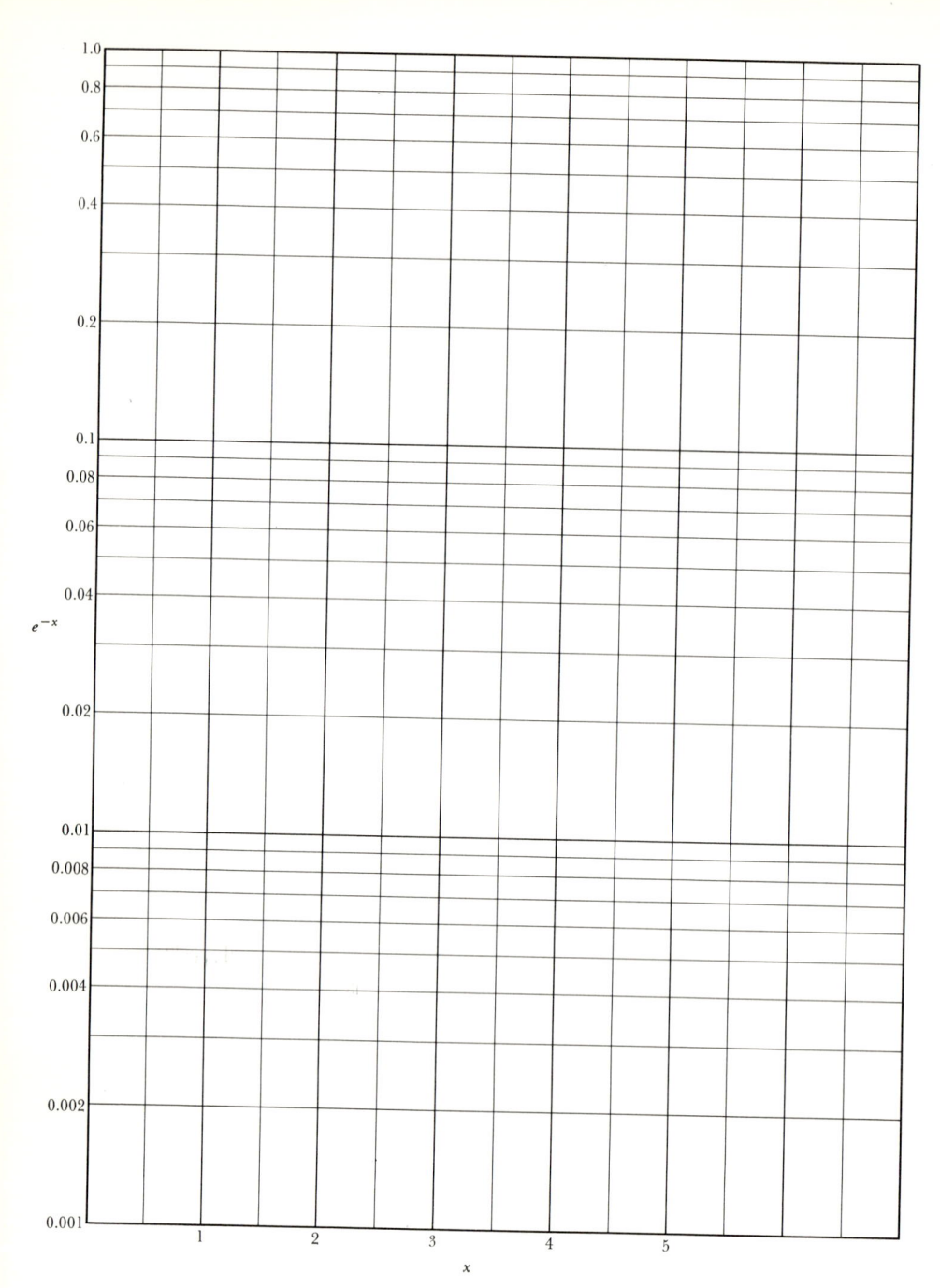

FIG. A.2. *A* **do-it-yourself** *plot of* e^{-x} *versus* x *on "semi-logarithmic" coordinates (see Problems A.1c and A.1d). I trust that your points fell on a straight line. When working problems throughout the book, approximate values of* e^{-x} *can be read from your graph as needed.*

Answers to Problems

Chapter One

1.1.	0.45 hours	*1.4.*	500; 250; 125; 125
1.2.	1.74	*1.5a.*	2
1.3b.	0.72 hours	*1.5b.*	2×10^{-8}
1.3c.	360	*1.6.*	About 3×10^{-8}
1.3d.	10^{-4}		

Chapter Two

2.1a.	111	*2.1g.*	267
2.1b.	126	*2.1h.*	251
2.1c.	151	*2.1i.*	6.6×10^6
2.1d.	135	*2.2.*	Not more than 68 μ
2.1e.	227	*2.3.*	15,000
2.1f.	242	*2.4.*	$10^{6.021}$

Chapter Three

3.1a.	2	*3.3a.*	1 min
3.1b.	2	*3.3b.*	2×10^5 pairs/min
3.1c.	6	*3.4.*	Adenine = uracil = 22.5 percent; guanine = cytosine = 27.5 percent
3.1d.	All		
3.2.	The "heavy" DNA is 4 percent more dense than the "light" DNA.	*3.5b.*	$\tfrac{2}{3}$

Chapter Four

4.1.	10^{-4}	*4.3a.*	One
4.2a.	None.	*4.3b(1).*	One

207

4.2b.	1/8	4.3b(2).	10^{-5}
4.2c.	7/16	4.3c	4×10^{-6}
4.2d.	1/2	4.4.	10 percent

Chapter Five

5.1a.	17 min	5.3b(4).	2, that is, a single
5.1b.	30–45 min	5.3c.	2 singles: 1 twin
5.2a.	11μ/min	5.3d.	Single
5.2b.	About 1/5 as long	5.3e.	2 : 1
5.3a.	2	5.3f(1).	All singles
5.3b(1).	0	5.3f(2).	10 : 1
5.3b(2).	4, that is, a twin	5.3f(3).	2 : 1
5.3b(3).	None		

Chapter Six

6.1a.	2	6.3.	37 ideal map units
6.1b.	20	6.5a.	All tetratype
6.1c.	1/2	6.5b(1).	2/3
6.1d.	1/4 ; 1/2	6.5b(2).	2/3; 2/3; 2/3; 0, 2/3; 0
6.2b(2).	0.22; 29 ideal map units	6.6a.	1 : 1
6.2c(2).	S = 0.45; double recombinant frequency = 0.020	6.6b.	1
		6.6c.	1/2; 3/4; 2/3
		6.6d.	2/3, 2/3

Chapter Seven

7.1b(1).	2.5	7.2.	About 1.7 Morgans; about 0.65 percent
7.1b(2).	2.1		
7.1b(3).	1.9		

Chapter Eight

8.1a.	10^{-6}	8.4b.	0.57
8.3a.	B	8.4c.	The transducing strain has 87 percent as much DNA as does λ.
8.3b.	ADC		
8.4a.	0.50		

Chapter Nine

9.1a.	4.1	9.1c.	Right
9.1b.	1.7	9.1d.	m_1

Chapter Ten

10.1.	4 nucleotides
10.2.	arg, glu; asn, thr, ileu, met, gln
10.3a.	met-thr-tyr-his-arg-pro-phe-gly

Answers to Problems 209

10.3b. met-thr-ileu-ileu-gly-pro-leu-gly
10.3c. met-thr-ser
10.3d. met-thr-ser-ser-ala-leu-try
10.3e. met-thr-ser-try-ala-leu-try;
met-thr-ser-gln-ala-leu-try
10.5a. Approximately 120.
10.5b(1). 4.0×10^{-7}
10.5b(2). 125

Chapter Eleven

11.1a. ½

11.1b.
Genotypes	Frequencies
A/A [1]	¼
A/a	½
a/a	¼

11.1c. ¼

11.1d.
Genotypes	Frequencies of types if loci 1 and 2 are unlinked	recombine 20% of the time
AB	¼	0.4
ab	¼	0.4
Ab	¼	0.1
aB	¼	0.1

11.1e.
Phenotypes	Frequencies of types if loci 1 and 2 are unlinked	recombine 20% of the time
AB	$\frac{9}{16}$	0.66
Ab	$\frac{3}{16}$	0.09
aB	$\frac{3}{16}$	0.09
ab	$\frac{1}{16}$	0.16

11.1f.
Genotypes	Frequency of type if S is	
	0.6	0
aBd	0.006	0
AbD	0.006	0

11.2a. 3
11.2b. 2-1-4
11.2c. Loci 2 and 1 recombine 40 percent of the time. Loci 1 and 4 recombine 10 percent of the time.
11.2d. 0.6

11.3a. 0.20
11.3b. 0.42; 0.01
11.4a. 0.01 and 0.20, respectively
11.4b. 2
11.5. ½

---

[1] The notation A/A means that each of the homologous Chromosomes in the diploid carries the allele A at locus 1.

Appendix

A.1d.	0.225; 0.027; 0.013; 0.74; 0.91	**A.3b(1).**	1.51×10^{-4}
A.2a.	$\frac{42}{90}$	**A.3b(2).**	0.063
A.2b.	$\frac{42}{90}$	**A.3b(3).**	0.368
A.2c.	$\frac{5}{12}$	**A.3c(1).**	$\frac{1}{4}; \frac{1}{2}$
A.2d.	$\frac{5}{12}$	**A.3c(2).**	$\frac{1}{8}; \frac{1}{4}$
A.2e.	$\frac{3}{10}$	**A.3c(3).**	$\frac{5}{16}$
A.3a(1).	0.49	**A.3c(4).**	$\frac{1215}{4096}$
A.3a(2).	0.32	**A.3d(1).**	0.368
A.3a(3).	$\frac{N!}{(N-n)!n!}(0.7)^n(0.3)^{N-n}$	**A.3d(2).**	0.061
		A.3d(3).	0.264

Subject Index

Acridine dyes, in mutation induction, 186-188
Additions as frame-shift mutations, 188
Adenine:
  deamination, 63
  in DNA, 33
  in viral RNA, 40
Allele, definition, 94
Amino acids:
  codons for, 189
  in peptide chains, 7-8
  structural formulas, 5
Anaphase:
  meiotic, 98
  mitotic, 85
Annealing, of nucleic acids, 185
Ascospore formation, 167-168
Ascus, 168
ATP, in energy conversion, 7
Auxotroph, definition, 11

Bacillus subtilis:
  DNA duplication, 88-90
  transformation, 21, 90
Bactereophage (*see* Phage)
Bacteria:
  chemical characterization, 7
  chromosome structure, 78-81
  conjugation, 78-80
  DNA duplication, 43-47, 88-90
  DNA length, 76
  enzymes, 7
  episomes, 161-162
  growth equation, 4
  growth medium, 5
  heteroduplexes, 66-67
  heterozygotes, 195-196
  Mendelian units in, 3
  mutation, 7-14
  partial diploids, 195-196
  in phage assay, 60
  phage growth in, 23-24
  phenotypic lag, 193-194
  transformation, 19-23
Binomial distribution, definition, 201
Biotin, transduction, 155-158
Bivalent, definition, 98
5-Bromouracil, induction of mutation, 62-65

Centromere, 83

Chiasma, definition, 98
Chlamydomonas:
  diploidy in, 196
  life cycle, 98-99
  meiosis in, 95, 98-99
Chromatid, definition, 83
Chromosomes:
  of bacteria, 78-81, 162
  duplication, 88-91
  of higher organisms, 81-88
  of phage, 138-139, 147-149
  of viruses, 76-78, 80-81
Cistron:
  definition, 143
  transcription, 185-186
Clone, definition, 9
Code, dictionary for, 189
Codons, definition, 188
Coefficient of coincidence:
  definition, 107
  in phage crosses, 134-135
Colchicine, 86
Complementation, definition, 194
Complementation test, definition, 143
Conjugation, in bacteria, 78-80, 161-162
Crossing over:
  definition, 101-102
  digressive, 102
  model for, 173-179
  progressive, 102
  regressive, 102
  sister chromatid, 102
Cytosine:
  deamination, 63
  in DNA, 33
  in viral RNA, 40

Deletions:
  definition, 67-68
  as frame-shift mutations, 188
  in heterozygotes, 147
  nitrous acid induction, 68
  reversion, 68
  in *rII* region, 144
Deoxyribonucleic acid (*see* DNA)
Deoxyribonucleotides:
  definition, 27
  neighbor relations in DNA, 33-37
  rare, 37-38
  structural formulas, 27
Deoxyribose, 27

Diakinesis, 98
Diplococcus pneumoniae (see Pneumococcus)
Diploid, definition, 95
Diplophase, definition, 196
DNA:
 annealing, 147
 in bacteria, 76
 duplication, 43-49, 50, 66-67, 86-91, 95, 137-138, 162
 in *Drosophila*, 76
 infectious, 27
 mutation, 54-70
 in phage, 23-24
 recombinant structure, 139
 single chain, 33, 37
 structure, 27-36
 in transformation (see also Transformation), 23
Dominance, definition, 194
Drosophila, DNA length, 76

Enzymes:
 altered by mutation, 14
 definition, 7
 in recombination, 141-142
 "repair," 142
Episome, definition, 160
Escherichia coli (see Bacteria)

Fertilization, 95
F factor:
 in bacterial fertility, 161-162
 incorporation, 162
 modified, 161
Fixed pairing segments, 171-173

Galactose:
 from hydrolysis of lactose, 15
 transduction, 155-158
β-Galactosidase, definition, 15
Gamete, definition, 95
Gene, definition, 144
Genotype, definition, 193
Guanine:
 deamination, 63
 in DNA, 33
 in viral RNA, 40

Haploid, definition, 95
Heteroduplex:
 definition, 58
 mutational, 65-67
 recombinational, 140, 147, 171
 segregation, 66-67
Heterozygotes:
 in bacteria, 195-196
 in phage, 139, 147
Hfr, definition, 161
Histones, 81
Homozygosity:
 definition, 196
 by inbreeding, 196

Hybridization experiments, 196
Hydrogen bond, definition, 32
Hydroxylamine, 65
Hypoxanthine, 63

Immunity, definition, 155
Interference:
 chromatid, 102
 definition, 107
 "high negative," 139, 141-142, 168-169
 negative, 132
 positive, 173
Interphase, 84

Jehle model for DNA duplication (see Jehle, H., *in Author Index*)

Lactose, hydrolysis, 15
λ phage:
 deletions, 67
 DNA content, 68
 DNA structure, 149-151
 linkage map, 131-139
 prophage locus, 155
 role in growth of *rII* mutants, 194
 transduction, 155-158
Leptotene, 95
Linkage map:
 circular, 138-139, 146
 definition, 101
 of λ*dg*, 156
 phage, 131-139
 topological, 145-146
Locus, definition, 94
Lysogenization, definition, 155
Lysozyme, mutants of phage, 195

Map distance, definition, 104
Mapping function:
 in absence of interference, 102-105
 phage, 131-139
Marker, definition, 94
Meiosis:
 generalized description, 95-98
 in salamander spermatocytes, 112-130
Metaphase:
 meiotic, 98
 mitotic, 85
6-Methylaminopurine, 38
Microorganisms (see Bacteria, Phage, and Virus)
Mitosis:
 generalized description, 82-86
 postmeiotic, 168
mRNA:
 definition, 182
 molecular weight, 186
Mutation:
 acridine-induced, 186-188
 in bacterial cultures, 7-14
 5-bromouracil-induced, 62-65
 deletions, 67
 effect on enzymes, 14

Subject Index 213

Mutation (Cont.)
 frame-shift, 186-188, 190
 heteroduplexes, 65-67
 hydroxylamine-induced, 65
 by ionization, 58-59
 multisite, definition, 144
 mutagenic, 68-69, 190
 nitrous acid-induced, 62-65
 phage-induced, 161
 by tautomerization of bases, 54-58
 transitions, 58-65
 transversion, 58, 190
Mutation rate:
 definition, 11
 measurement, 11-14
 in phage, 61

Nitrogenous bases (see also Deoxyribonucleotides, DNA, RNA, Thymine, Uracil, Cytosine, Guanine, Adenine:
 in DNA, 27-36
 in the genetic code, 189
 in nucleotides, 27
 tautomerism, 54-58
 in viral RNA, 39-40
Nitrous acid, induction of mutation, 62-65
Neurospora crassa, life cycle, 167-168
Nuclear fusion, definition, 95

P22 (phage):
 permuted chromosome, 160
 transduction, 160
P32, inactivation of phage, 67-68
Pachytene, 98
Penicillin, 15
Penicillinase, 15
Peptide chain, definition, 7-8
Phage:
 assay, 59
 chemical composition, 23
 chromosome, 77-78, 147-149
 complementation, 195
 cross:
  definition, 131
  triparental, 132
 DNA, 23-24, 37, 74
 heteroduplexes, 65
 heterozygotes, 139
 5-hydroxymethyl cytosine in, 37
 life cycle, 23-24, 78
 maturation, 78
 mutation rate, 61
 mutator, 161
 particle dimensions, 74
 phenotypic mixing, 194
 recombination, 131-153, 174
 temperate, definition, 154
 virulent, definition, 154
Phenotype, definition, 193
Phenotypic lag, 193-194
Phenotypic mixing, 193-194

øx-174 (phage), transcription, 185
Plaque:
 mottled, 65
 phage assay method, 60
Pneumococcus, transformation, 19-23
Poisson distribution:
 of crossovers, 104-105
 definition, 201-203
 in measurement of mutation rate, 12
 of phage on host cells, 132
Polycytidilic acid, 65
Polyuridylic acid, 189
Prophage, definition, 155
Prophase:
 meiotic, 95
 mitotic, 84
Protein:
 in Chromosomes, 81
 definition, 7
 synthesis, 182-186, 188-189
Prototroph, definition, 11
Purine (see Nitrogenous bases)
Pyrimidine (see Nitrogenous bases)

Qβ (phage), in vitro RNA duplication, 50

rII:
 deletions, 68, 144
 dominance-recessive relations, 194
 genetic code in, 186-188
 mutations, 59-62, 65, 67
 plaque morphology, 60
 recombination, 143-146
 reversion rates, 143
Recessiveness, definition, 194
Recombination:
 assymetric, 171
 in bacteria, 154-166
 deficient mutants, 163
 definition, 95
 enzymes, 141-142
 in higher organisms, 94-130, 167-180
 mechanism, 163
 in phage, 131-153
 nonreciprocal, 171, 174
 in rII region, 61
 reciprocal, 174
Replica plating, 9
Ribosomes, definition, 182
RNA:
 in Chromosomes, 81
 infectious, 27
 as genetic material, 27
 polymerase, 65
 synthesis, 50, 65, 185-186
 viral, 27, 39-40

Salamanders, meiosis in, 112-130
Segregation:
 aberrant in ascomycetes, 169-171
 meiotic, 99-100
 postmeiotic, 171

Spindle:
  meiotic, 98
  mitotic, 83
sRNA (see tRNA)
"Suicide" (see P^{32} inactivation of phage)
Synopsis, definition, 98

T2 (see Phage)
Tautomerization (see Mutation, by tautomerization of bases)
Telophase:
  meiotic, 98
  mitotic, 85
Terminalization, 98
Terminal redundancy, 147-151
Tetrad:
  definition, 100
  parental ditype, 100
  recombinant ditype, 100
  tetratype, 100
Thymine, 33
Tobacco mosaic virus (TMV), 27
Transaminase, 14
Transcription, definition, 182
Transduction, 155-158; 160-161
Transformation:
  in *B. subtilis*, 90
  general properties, 19-23
  mechanism, 163
Transitions, 58

Translation, definition, 182
Transversions, 58, 190
tRNA, definition, 183

Ultraviolet light, 65
Uracil:
  in DNA, 38
  in viral RNA, 40

Vicia faba, 86
Virus (see also Phage):
  chromosomes, 76-78
  DNA, 33, 91
  RNA, 27, 50

Watson-Crick model for DNA duplication (see DNA, duplication)
Watson-Crick model for DNA structure (see DNA, structure)
Watson-Crick model for mutation (see Mutation, by tautomerization of bases)
Wild-type, definition, 9

Xanthine, 63

Yeast, 171, 173

Zygotene, 98

Author Index

Numbers in italic are pages on which complete references appear.

Adelberg, E. A., *16, 40*
Amati, P., *134, 135, 151*
Avery, O. T., 22, *23*, *40*

Barratt, R. W., *108*
Benzer, S., 61, 143, *151*
Bishop, D. H. L., *51*
Bonner, D. M., 20
Buttin, G., *164*

Cairns, J., 47, *51*, 77, 88
Calef, E., 157, *164*
Campbell, A., 157, 158, *164*
Chase, M. C., 23, *41*
Clark, A. J., *164*
Claybrook, J. R., *51*
Colin, M., *40*
Creighton, H. B., *108*
Crick, F. H. C., 33, *41*, 42, 46, *51*, 54, 58, 61, 63, *70*, 186, 187, 188
Cuzin, F., *164*

Delbrück, M., 14, *16*, 18, 133
De Robertis, E. D. P., *91*
Dintzes, H., 186

Feller, W., *203*
Fogel, S., 171, 173, *179*
Fraenkel-Conrat, H. L., *40*
Franklin, R. E., 33
Freese, E., *69*

Garnjobst, L., *108*
Gosling, R. G., 33
Goulian, M., *51*
Griffith, F., 19

Hartman, P. E., 185, 186
Haruna, I., *51*
Hershey, A. D., 23, *24, 41*
Herskowitz, I. H., *185*
Hodgman, C. D., *203*
Hogness, D., *136*
Holliday, R., *179*
Horne, R. W., *72*
Huberman, J., *150*
Hurst, D. D., 171, 173, *179*

Jacob, F., *78, 80, 91*, 160, *164*
Jehle, H., 46, 47, *51*, 66, 67, *70*, 140
Jinks, J. L., 160

Kaiser, D., *136*
Karam, J. D., *69*
Kezer, J., *113*
Kleinschmidt, A. K., *79*
Kornberg, A., 33, *51*, 52, 53, 63, 142
Kreig, D. R., *69*

Lederberg, E. M., *16*
Lederberg, J., *16*
Lenny, A. B., *69*
Levine, P., *98*
Licciardello, G., 157, *164*
Luria, S. E., 14, *16*, 18, 61, *69*

McCarty, M., 22, *23, 40*
McClintock, B., *108*
McElroy, W. D., *6, 16*
MacHattie, L., 149
MacLeod, C. M., 22, *23, 40*
Mason, D. J., 82
Mendel, G., *196*
Merz, T., 102, *108*
Meselson, M. S., 43, *45*, *51*, 88, *134, 135, 136, 137, 151, 164, 175, 179*
Mills, D. R., *51*
Murray, N., 171, *179*

Newmeyer, D., *108*
Nirenberg, M., *189*
Nowinski, W. W., *91*

Pace, N. R., *51*
Paigen, K., *164*
Perkins, D. D., *108*
Peters, J. A., *40, 41, 108*
Peterson, R., *51*

Robley, C. W., *40*

Saez, F. A., *91*
Sinsheimer, R. L., *51*
Speyer, J. F., *69*
Spiegelman, S., 50, *51*
Stadler, D. R., 172
Stahl, F. W., *51, 69, 203*
Stent, G. S., *41, 69*
Streisinger, G., *69*, 190, 191
Sturtevant, A. H., *108*
Sueoka, N., 90

Suskind, S. E., 185, 186
Swanson, C. P., 102, *108*

Tatum, E. L., *3*
Taylor, J. H., 86, *87*, 88, *91*, 93
Terzaghi, B. E., *69*
Thomas, C. A., Jr., *91*, 147, 149
Tomizawa, J. I., *140*

Visconti, N., 133

Watson, J. D., 33, *41*, 42, 46, *51*, 53, 54, 58, 61, 63, 70
Weigle, J. J., *136, 137, 151, 164*
Whitehouse, H. L. K., 179
Wilkins, M., 33, *41*, *83*
Wollman, E. L., *78, 80, 91*, 160, 161, *164*

Yoshikawa, H., 90
Young, W. J., 102, *108*